Biodegradable Waste Management

Biodegradable Waste Management

Lydia Hemsworth

Larsen & Keller
www.larsen-keller.com

Biodegradable Waste Management
Lydia Hemsworth
ISBN: 978-1-64172-419-7 (Hardback)

Larsen & Keller
Published by Larsen and Keller Education,
5 Penn Plaza,
19th Floor,
New York, NY 10001, USA

Cataloging-in-Publication Data

Biodegradable waste management / Lydia Hemsworth.
p. cm.
Includes bibliographical references and index.
ISBN 978-1-64172-419-7
1. Refuse and refuse disposal. 2. Materials--Biodegradation.
3. Biodegradable products. I. Hemsworth, Lydia.
TD791 .B56 2020
363.728--dc23

Printed and bound in China.

For more information regarding Larsen and Keller Education and its products, please visit the publisher's website www.larsen-keller.com

Table of Contents

Preface

This book has been written, keeping in view that students want more practical information. Thus, my aim has been to make it as comprehensive as possible for the readers. I would like to extend my thanks to my family and co-workers for their knowledge, support and encouragement all along.

Any organic matter in waste which can be broken down into water, methane, carbon dioxide and simpler organic molecules is known as biodegradable waste. It is broken down by processes like aerobic digestion, composting and anaerobic digestion. Biodegradable waste includes manure, human waste, sewage and slaughterhouse waste. Food waste, green waste, paper waste and biodegradable plastics also fall under this category. Biodegradable waste management involves the treatment of biodegradable waste by processes such as pyrolysis, biodegradation and bioremediation. Pyrolysis involves the incineration of dry waste matter in order to produce heat and electricity. This book is compiled in such a manner, that it will provide in-depth knowledge about the theory and practice of biodegradable waste management. Most of the topics introduced in it cover new techniques of waste management. This textbook is an essential guide for both academicians and those who wish to pursue this discipline further.

A brief description of the chapters is provided below for further understanding:

Chapter - What is Biodegradable Waste?

The organic matter in waste that can be broken down into carbon dioxide, methane, water and simple organic molecules with the action of microorganisms is referred to as biodegradable waste. This is an introductory chapter which will provide a brief introduction to the major aspects of biodegradable waste and its management.

Chapter - Sources of Biodegradable Waste

The main sources of biodegradable waste are slaughterhouses, food, feces, biodegradable plastics and sewage sludge. Biodegradable waste also comes from municipal soil waste and organic waste. This chapter has been carefully written to provide an easy understanding of these sources of biodegradable waste along with the recycling of organic waste.

Chapter - Types of Biodegradable Waste Management

Biodegradable waste management is divided into diverse types such as slaughterhouse waste management, municipal solid waste management, food waste management and sewage sludge treatment. The topics elaborated in this chapter will help in gaining a better perspective about these diverse kinds of waste management.

Chapter - Processes for Biodegradable Waste Management

Some of the major concepts and processes related to biodegradable waste management are decomposition, aerobic digestion, anaerobic digestion, pyrolysis, bioremediation, biodegradation, etc. The chapter closely examines these key concepts and processes of biodegradable waste management.

Chapter - Waste Handling and Disposal

The process of collecting and transporting waste is termed as waste handling. There are numerous methods which are used to dispose waste such as incineration, waste compaction, biogas generation and composting. All the diverse techniques and methods of waste handling and disposal have been carefully analyzed in this chapter.

Lydia Hemsworth

1

What is Biodegradable Waste?

The organic matter in waste that can be broken down into carbon dioxide, methane, water and simple organic molecules with the action of microorganisms is referred to as biodegradable waste. This is an introductory chapter which will provide a brief introduction to the major aspects of biodegradable waste and its management.

WASTE

A natural part of the life cycle, waste occurs when any organism returns substances to the environment. Living things take in raw materials and excrete wastes that are recycled by other living organisms. However, humans produce an additional flow of material residues that would overload the capacity of natural recycling processes, so these wastes must be managed in order to reduce their effect on our aesthetics, health, or the environment.

Solid and fluid, hazardous and non-toxic wastes are generated in our households, offices, schools, hospitals, and industries. No society is immune from day-to-day issues associated with waste disposal. How waste is handled often depends on its source and characteristics, as well as any local, state, and federal regulations that govern its management. Practices generally differ for residences and industries, in urban and rural areas, and for developed and developing countries.

Municipal Solid Waste

Waste collected from residences, commercial buildings, institutions such as hospitals and schools, and light industrial operations is most often categorized as municipal solid waste. MSW consists primarily of paper, containers and packaging, food wastes, yard trimmings, and other inorganic wastes. Municipal solid waste can also include industrial sludge, classified as hazardous or non-hazardous, resulting from a wide array of mining, construction, and manufacturing processes.

In 2006, Americans generated more than 250 million tons of trash. Nearly 33 percent, 82 million tons of materials, was recycled; the energy equivalent of more than 10 billion gallons of gasoline. More than 30 million tons (12.5 percent) were combusted through an energy recovery process, and approximately 138 million tons (55 percent) of materials were discarded in landfills. Municipal waste, when properly managed, does not pose an immediate threat to human health or the environment.

Hazardous Waste

Waste material that is flammable, corrosive, reactive, or toxic—which can be in the form of a solid, liquid, or gas—is defined as hazardous waste. Although the term often evokes an image of items marked with skull and crossbones, many hazardous wastes include products used every day, including paint, used oil from cars, batteries, shoe polish, and even laundry detergent. In addition, many of the items that we rely upon generate hazardous waste during the process of their production.

The U.S. Environmental Protection Agency (EPA) reported that 279 million tons of hazardous wastes were generated in 1996; ninety-six percent of which was industrial process water by-product waste. In 1997 the EPA made a rule change that separated industrial wastewater from hazardous waste reporting. This change can clearly be seen in current reporting numbers; in 2005 the EPA reported hazardous waste generation of just under 38.5 million tons.

Businesses that generate hazardous wastes are required by legislation to manage them from generation to disposal. The waste is often treated to change its biological, chemical, or physical characteristics in order to make it less hazardous or to reduce its overall volume. Some hazardous materials can be recycled if it is environmentally safe to do so, although it can be expensive. Any leftover waste is then safely disposed of to further neutralize any adverse affects to human health or the environment. Today, many industries are attempting to reduce their generation of hazardous waste by modifying their manufacturing processes or by replacing hazardous materials with less hazardous or non-hazardous substitutes.

E-Waste

A new and growing segment of our waste stream is termed? e-waste. Although not clearly defined, e-waste applies to much of the electronic equipment used by businesses and individual consumers that are nearing the end of their usefulness. This includes, but is not limited to, computers, fax machines, copiers, and televisions.

A factor that complicates disposal of these items is that certain components contain hazardous materials. The cathode ray tubes in computer monitors and televisions are an example. As such, many old electronics sit idle due to the uncertainty of how to manage them.

Yet, many of these products can be reused, refurbished, or recycled. There are a number of options to reuse equipment that can still function—from selling it to someone who can use it or giving it as a charitable donation. Many non-working items can be refurbished in order to recover them into working condition. Finally, any components that cannot be repaired can often be recycled.

Although there no federal regulation for e-waste, some states have taken various approaches toward its management. Many states, including Massachusetts, Florida, and New York, have streamlined their regulations to increase the level of recycling. On the other hand, in 1993 California passed the Electronic Waste Recycling Act; yet, they also regulate cathode ray tubes as hazardous waste, banning them from regular trash disposal.

BIODEGRADABLE WASTE

Biodegradable waste is a type of waste, typically originating from plant or animal sources, which may be degraded by other living organisms. Waste that cannot be broken down by other living organisms are called non-biodegradable.

Biodegradable waste can be commonly found in municipal solid waste (sometimes called biodegradable municipal waste, or BMW) as green waste, food waste, paper waste, and biodegradable plastics. Other biodegradable wastes include human waste, manure, sewage, slaughterhouse waste. In the absence of oxygen much of this waste will decay to methane by anaerobic digestion.

Climate Change impacts

The main environmental threat from biodegradable waste is the production of methane in landfills. Methane is 21 times as potent a greenhouse gas as carbon dioxide and accounted for some 3% of total greenhouse gas emissions in the EU-15 in 1995. The Landfill Directive 1999/31/EC obliges Member States to reduce the amount of biodegradable waste that they landfill to 35% of 1995 levels by 2016, which will significantly reduce the problem.

Uses of Biodegradable Waste

Biodegradable waste can often be used for composting which turns the waste into humus. It can also be used as a resource for heat, electricity and fuel in future, by using anaerobic digestion as it is being achieved by the swiss Kompogas treatment for 20 years now. This produces additional Biogas and still delivers usable nutrients that can be implemented to the soil. Especially animal-based kitchen scraps (such as leftover fish, meat) are best disposed this way (as these are best not composted). Finally note that leftover human food (vegetables -no animal flesh-) can also be used as fodder to flightless birds, goats, pigs, sheep or alternatively can be fed to birds, and other wildlife.

In the USA, approximately 50% of waste sent to landfills or incinerators is biodegradable; therefore, composting could create many new jobs, and recover significant resources that are being lost.

DIFFERENCE BETWEEN BIODEGRADABLE AND NON-BIODEGRADABLE WASTE

Biodegradable substances are those that degrades or break down naturally. Non-biodegradable substances are those that do not degrade easily. These terms itself defines the ability of the substances which are degradable or not. Materials like plants, animals, their waste, paper, fruits, flowers, vegetables fall under biodegradable substances, on the other hand, rubber, plastic, chemicals, paint plastic falls under the category of the non-biodegradable items.

The environment in which we live is surrounded by these two things only, even the materials we use in day to day life can be biodegradable or non-biodegradable. It is usually seen that non-biodegradable substances are synthesized and does not occur naturally, hence their degradation is also not possible, but the biodegradable substances get easily decompose in nature through agents like water, soil, sunlight, microorganisms, etc. and directly or indirectly are profitable to the environment.

Comparison Chart

Basis For Comparison	Biodegradable Substances	Non-Biodegradable Substances
Meaning	The wastes that decompose naturally in the environment and are considered safe for the environment are called as biodegradable substances.	Such wastes which do not decompose naturally in the environment causes pollution and are also harmful to the living being are called the non-biodegradable substances.
Examples	Dead plants and animals, their waste, fruits, vegetables, flowers, paper, etc.	Chemicals, paints, plastic, rubber, toxic, plastic, metals, etc.
The rate of decomposition	Slow.	Fast.
Decomposed by	Bacteria, fungi, and other living organisms have the capability of decomposing the material into the soil.	These substances do not degrade easily or by the action of natural agents.
Use	After degradation, they can be used to produce biogas, manure, fertilisers, compost.	As their degradation rate is slow and the separation and recycling are not easy and expensive too.
	Biodegradable substances are not harmful to the environment.	They are harmful to the environment.

Biodegradable

The degradation or decomposition refers to the substances that get decay with the help of natural agent such as microorganisms, air, water, soil, sunlight, ozone, etc. into the organic manure. Further, these organic matters are profitable to the environment in some, or the other way is known as biodegradable substances.

These materials are non-toxic to the environment and mainly include the natural substances only. Plants and animals waste, even the dead plants and animals, fruits, paper, vegetables, etc. get convert into the simpler units, which further get into the soil and are used as manures, biogas, fertilizers, compost, etc. They are non-toxic, non-pollutants, eco-friendly to the environment.

Nowadays, an emphasis is given to using biodegradable and eco-friendly items as much as possible to reduce the level of pollution and substance responsible for it. This will also contribute to reducing the pollution up to some level, which is becoming one of the major problems worldwide and is hazardous to every living being.

Non-biodegradable

Materials that remain for a long time in the environment, without getting decompose by any natural agents, also causing harm to the environment are called non-biodegradable substances. These materials are metals, plastics, bottles, glass, poly bags, chemicals, batteries, etc.

But as these are readily available, convenient to use, and are of low cost, the non-biodegradable substances are more often used. But instead of returning to the environment, they become solid waste which cannot be broken down and become hazardous to the health and the environment. Hence are regarded as toxic, pollution causing and are not considered as eco-friendly.

Many measures are taken these days, concerning the use of non-biodegradable materials. The three 'R' concept which says Reduce-Recycle -Reuse is in trend, which explains the use of the non-biodegradable materials. As we already discuss that these substances do not decompose or dissolve easily so can be recycled and reuse. And one can help in reducing this waste by instead of throwing the plastics and poly bags in the garbage; it can be put in the recycling bags to use again.

2

Sources of Biodegradable Waste

The main sources of biodegradable waste are slaughterhouses, food, feces, biodegradable plastics and sewage sludge. Biodegradable waste also comes from municipal soil waste and organic waste. This chapter has been carefully written to provide an easy understanding of these sources of biodegradable waste along with the recycling of organic waste.

SLAUGHTERHOUSE WASTE

Slaughterhouse wastes are a potential reservoir of bacterial, viral, prion and parasitic pathogens, capable of infecting both animals and humans. A quick, cost effective and safe disposal method is thus essential in order to reduce the risk of disease following animal slaughter. Different methods for the disposal of such wastes exist, including composting, anaerobic digestion (AD), alkaline hydrolysis (AH), rendering, incineration and burning. Composting is a disposal method that allows a recycling of the slaughterhouse waste nutrients back into the earth. The high fat and protein content of slaughterhouse wastes mean however, that such wastes are an excellent substrate for AD processes, resulting in both the disposal of wastes, a recycling of nutrients (soil amendment with sludge), and in methane production. Concerns exist as to whether AD and composting processes can inactivate pathogens. In contrast, AH is capable of the inactivation of almost all known microorganisms.

Slaughterhouse wastes are animal byproducts, and can contain different bacterial, viral, prion and parasitic pathogens. Composting and AH were also reviewed, with regard to their abilities to eliminate specific pathogens.

Slaughterhouse wastes constitute the inedible parts of animals derived from the production of meat, as well as blood and other animal byproducts. Inedible animal tissues (organs, integument, ligaments, tendons, blood vessels, feathers, bone) can comprise up to 45% or more of the slaughtered animal. The remaining fraction, is, however, not only fit for disposal, and pet food companies are known to purchase large amounts of slaughterhouse wastes. These can be used either alone, or as a supplement for animal feed. Nonetheless, considerable amounts of slaughterhouse waste are produced worldwide, and their disposal poses a serious logistical challenge for meat/poultry processing plants. As a result of such high magnitudes of animal waste, and because of the legal restrictions and rising treatment costs of removal, improper and unsafe disposal of these wastes can occur. Such practices can then also lead to serious environmental problems. Slaughterhouse wastes have, however, potential as an energy source, and could help to reduce the requirements of petroleum based fuels currently used for the Earth's energy needs.

As such alternative energy sources emerge, a number of questions will be asked concerning the safety of the end products generated. In the case of using slaughterhouse wastes in biogas production, the microbial quality of the end product is of major concern as slaughterhouse wastes can be contaminated with high numbers of microorganisms including bacteria, viruses, prions, fungi, yeasts and associated microbial toxins. Such wastes pose a potential risk to animal and human health, unless handled and treated properly.

The management of animal carcasses has always been, and continues to be, a concern in animal production operations, slaughter plants, farms and other facilities that involve animals. Throughout history, burial and to a lesser extent, burning, have been the most commonly applied methods for the disposal of on-farm mortalities. However, the European Union Animal By-Product Regulation (EC) no. 1774/2002 does not allow these practices to be conducted within the EU and limits the disposal routes to incineration (either on or off-farm), rendering, high temperature/pressure AH, disposal at maggot farms or through licensed waste collectors. The prevention of disposal by burial and burning was founded based on the perceived risk of an incomplete destruction of pathogens from mortalities during these processes, and thus the entering of infective agents into the animal feed chain. Similar risks are involved upon the slaughtering of animals. In contrast to burning, properly operated incineration facilities pose fewer pollution concerns. Also, bacteria (including spore-formers) and viruses should not survive the incineration process. There are, however, some concerns that transmissible spongiform encaphalopathies (TSEs) such as BSE (bovine spongiform encephalopathy) can survive incineration processes if not conducted at a high enough temperature.

Rendering is another method used for the disposal of animal waste products. It involves the conversion of animal carcass/waste materials into three end products, carcass meal (proteinaceous solids), melted fat/tallow, and water. This is conducted using mechanical processes including grinding, mixing, pressing, decanting and separating, thermal processes including cooking, evaporating, and drying, as well as chemical processes such as solvent extraction. The rendering process simultaneously dries the material and separates the fat from the bone and protein. The fat obtained can be used as low-cost raw material in making grease, animal feed, soap, candles and biodiesel, and tallow is an important raw material in the steel rolling industry providing the required lubrication for compressing steel sheets. The protein meal produced can be used for animal feed. Thus the products provide significant additional income to the slaughterhouse. However, because of the problems with BSE, the feeding of meat and bone meal to cattle is currently prohibited in developed countries. Thus rendering plants do not play as significant a role in the disposal of animal wastes today as they did in the past.

New risks have arisen from the implementation of the European Union (EU) Landfill Directive for more environmentally sensitive waste disposal methods. Evidence exists that certain pathogens and pests can survive composting or other waste treatment processes, sometimes through inadequate methods or failures in the process. According to the European Union Animal By-Product Regulation, animal byproducts (ABP) are categorized in three classes. ABP of categories can be treated by approved composting plants and by AD in approved biogas plants, while ABP one of category cannot.

MUNICIPAL SOLID WASTE

Municipal solid waste (MSW), commonly known as trash or garbage in the United States and rubbish in Britain, is a waste type consisting of everyday items that are discarded by the public. "Garbage" can also refer specifically to food waste, as in a garbage disposal; the two are sometimes collected separately. In the European Union, the semantic definition is 'mixed municipal waste,' given waste code 20 03 01 in the European Waste Catalog. Although the waste may originate from a number of sources that has nothing to do with a municipality, the traditional role of municipalities in collecting and managing these kinds of waste have produced the particular etymology 'municipal.'

Composition

The composition of municipal solid waste varies greatly from municipality to municipality, and it changes significantly with time. In municipalities which have a well developed waste recycling system, the waste stream mainly consists of intractable wastes such as plastic film and non-recyclable packaging materials. At the start of the 20th century, the majority of domestic waste (53%) in the UK consisted of coal ash from open fires. In developed areas without significant recycling activity it predominantly includes food wastes, market wastes, yard wastes, plastic containers and product packaging materials, and other miscellaneous solid wastes from residential, commercial, institutional, and industrial sources. Most definitions of municipal solid waste do not include industrial wastes, agricultural wastes, medical waste, radioactive waste or sewage sludge. Waste collection is performed by the municipality within a given area. The term *residual waste* relates to waste left from household sources containing materials that have not been separated out or sent for processing. Waste can be classified in several ways but the following list represents a typical classification:

- Biodegradable waste: Food and kitchen waste, green waste, paper (most can be recycled although some difficult to compost plant material may be excluded).
- Recyclable materials: Paper, cardboard, glass, bottles, jars, tin cans, aluminum cans, aluminum foil, metals, certain plastics, fabrics, clothes, tires, batteries, etc.
- Inert waste: Construction and demolition waste, dirt, rocks, debris.
- Electrical and electronic waste (WEEE): Electrical appliances, light bulbs, washing machines, TVs, computers, screens, mobile phones, alarm clocks, watches, etc.
- Composite wastes: Waste clothing, Tetra Packs, waste plastics such as toys.
- Hazardous waste including most paints, chemicals, tires, batteries, light bulbs, electrical appliances, fluorescent lamps, aerosol spray cans, and fertilizers.
- Toxic waste including pesticides, herbicides, and fungicides.
- Biomedical waste, expired pharmaceutical drugs, etc.

Components of Solid Waste Management

The municipal solid waste industry has four components: recycling, composting, disposal, and waste-to-energy via incineration. There is no single approach that can be applied to the management of all waste streams, therefore the Environmental Protection Agency, a U.S. federal government agency, developed a hierarchy ranking strategy for municipal solid waste. The Waste Management Hierarchy is made up of four levels ordered from most preferred to least preferred methods based on their environmental soundness: Source reduction and reuse; recycling or composting; energy recovery; treatment and disposal.

Bins to collect paper, aluminium, glass, PET bottles and incinerable waste.

Collection

The functional element of collection includes not only the gathering of solid waste and recyclable materials, but also the transport of these materials, after collection, to the location where the collection vehicle is emptied. This location may be a materials processing facility, a transfer station or a landfill disposal site.

Waste Handling and Separation, Storage and Processing at the Source

Waste handling and separation involves activities associated with waste management until the waste is placed in storage containers for collection. Handling also encompasses the movement of loaded containers to the point of collection. Separating different types of waste components is an important step in the handling and storage of solid waste at the source of collection.

Segregation, Processing and Transformation of Solid Wastes

The types of means and facilities that are now used for the recovery of waste materials that have been separated at the source include curbside ('kerbside' in the UK) collection, drop-off and buy-back centers. The separation and processing of wastes that have been separated at the source and the separation of commingled wastes usually occur at a materials recovery facility, transfer stations, combustion facilities and treatment plants.

Transfer and Transport

This element involves two main steps. First, the waste is transferred from a smaller collection vehicle to larger transport equipment. The waste is then transported, usually over long distances, to a processing or disposal site.

Disposal

Mixed municipal waste, Hiriya, Tel Aviv.

Today, the disposal of wastes by land filling or land spreading is the ultimate fate of all solid wastes, whether they are residential wastes collected and transported directly to a landfill site, residual materials from materials recovery facilities (MRFs), residue from the combustion of solid waste, compost, or other substances from various solid waste processing facilities. A modern sanitary landfill is not a dump; it is an engineered facility used for disposing of solid wastes on land without creating nuisances or hazards to public health or safety, such as the problems of insects and the contamination of ground water.

Reusing

In the recent years environmental organizations, such as Freegle or Freecycle Network, have been gaining popularity for their online reuse networks. These networks provide a worldwide online registry of unwanted items that would otherwise be thrown away, for individuals and nonprofits to reuse or recycle. Therefore, this free Internet-based service reduces landfill pollution and promotes the gift economy.

Landfills

Landfills are created by land dumping. Land dumping methods vary, most commonly it involves the mass dumping of waste into a designated area, usually a hole or sidehill. After the waste is dumped, it is then compacted by large machines. When the dumping cell is full, it is then "sealed" with a plastic sheet and covered in several feet of dirt. This is the primary method of dumping in the United States because of the low cost and abundance of unused land in North America. Landfills are regulated in the US by the Environmental Protection Agency, which enforces standards provided in the Resource Conservation Recovery Act, such as requiring liners and groundwater monitoring. This is because landfills pose the threat of pollution and can intoxicate ground water. The signs of pollution are effectively masked by disposal companies and it is often hard to see any evidence. Usually landfills are surrounded by large walls or fences hiding the mounds of debris.

Large amounts of chemical odor eliminating agent are sprayed in the air surrounding landfills to hide the evidence of the rotting waste inside the plant.

Energy Generation

Municipal solid waste can be used to generate energy. Several technologies have been developed that make the processing of MSW for energy generation cleaner and more economical than ever before, including landfill gas capture, combustion, pyrolysis, gasification, and plasma arc *gasification*. While older waste incineration plants emitted a lot of pollutants, recent regulatory changes and new technologies have significantly reduced this concern. United States Environmental Protection Agency (EPA) regulations in 1995 and 2000 under the Clean Air Act have succeeded in reducing emissions of dioxins from waste-to-energy facilities by more than 99 percent below 1990 levels, while mercury emissions have been reduced by over 90 percent. The EPA noted these improvements in 2003, citing waste-to-energy as a power source "with less environmental impact than almost any other source of electricity".

GREEN WASTE

Green waste is biodegradable waste that can be composed of garden or park waste. It includes things like grass clippings, shrub and yard clippings, branches, woodchips, bark, wood, palm trees and branches, and weeds. These are all rubbish materials you commonly encounter but you may not realise the differences between green and other types of waste.

The organic material which makes up green waste can be processed through a different disposal method. It's not like your other general household waste, or plastics. The green waste can be turned to a local processor, and turned to mulch or compost. Recycling green waste has a lot of benefits.

Green waste, also known as "biological waste," is any organic waste that can be composted. It is most usually composed of refuse from gardens such as grass clippings or leaves, and domestic or industrial kitchen wastes. Green waste does not include things such dried leaves, pine straw, or hay. Such materials are rich in carbon and considered "brown wastes," while green wastes contain high in concentrations of nitrogen. Green waste can be used to increase the efficiency of many composting operations and can be added to soil to sustain local nutrient cycling.

Collection of Green Waste

Green waste can be collected via municipal curbside collection schemes or through private waste management businesses. Many communities, especially in the United Kingdom, have initiated green waste recycling and collection programs in order to decrease the amount of biodegradable materials in landfills. Communities are provided with, or can provide their own, compost receptacles that they fill with plant and food remains, which are then emptied on a regular basis. Programs such as this allow communities to be an active part of composting their green waste which allows them to play an active role in decreasing the amount of food being dumped into local and regional landfills.

This is a bin filled with materials that comprise green waste, such as kitchen wastes and plant trimmings.

Uses

Green waste can be used to improve the quality and sustainability of industrially manufactured topsoils and the sanitariness and efficiency of sewage disposal.

Manufactured Topsoils

Green waste is an integral part of many manufactured topsoils, as it provides both nutrients for growing plants and increases the volume of manufactured topsoils. Its woody components do not decompose quickly, so they provide the bulk that is necessary for supplementary topsoils. Mixing industrial wastes such as fly ash or coal dust with green waste to create artificial topsoil not only facilitates the repurposing of industrial debris and keeps it out of landfills, but it also allows the nutrients in green waste materials to be cycled back into the environment. By utilizing fly ash in conjunction with green waste, manufactured soils are able to increase their water holding capacity while simultaneously recycling refuse materials that might otherwise take up space in a landfill. This allows fly ash and green waste to increase local soil nutrient levels and promotes natural nutrient cycling processes.

Sewage Disposal

Green waste can also be mixed with sewage wastes and composted, providing a safe, environmentally sustainable option for sewage disposal. Co-composting green and sewage wastes eliminates the risk that pathogens and pollutants contained in sewage wastes might pose to the environment. Utilization of green waste to dispose of sewage wastes not only decreases the amount of sewage incinerated and dumped each year, it also facilitates the cycling of organic wastes back into the environment. The co-composted remains of these organic wastes can be safely employed for use in agriculture. This process decreases the amount of trash being dumped into landfills and other trash repositories and allows for the complete cycling of organic nutrients through the environment.

Renewable Energy

Vegetable waste being dumped in a market in Hyderabad, India.

Biogas captured from biodegradable green waste can be used as biofuel. Green waste can be composed of non-food crops, which decompose to produce cellulosic ethanol. It can also help reduce the necessity of petroleum gases, which produce large amounts of greenhouse gases, such as carbon dioxide, when burned.

Soil Health

Green waste composting has also been linked to suppression of soil borne diseases such as damping off and root rots that affect large agricultural ventures like greenhouses and large-scale farms. This disease suppressive quality has positive implications for lesser-developed nations that do not have the technology or resources to purchase expensive fertilizers.

Addition of composts which contain residues and particles of heavy metals to soil can raise the soil's heavy metal content and increase the probability of these metals to transfer into crop plants. When biological, or green waste is added to these soil samples, plant uptake of heavy metal has been shown to decrease crop uptake of metals compared to other types of compost composed of things such as sewage sludge. This can protect consumers and the environment from biomagnification caused by long-term accumulation of heavy metal particles within the soil and plant life of an area.

FOOD WASTE

Food waste or food loss is food that is discarded or lost uneaten. The causes of food waste or loss are numerous and occur at the stages of producing, processing, retailing and consuming.

Global food loss and waste amount to between one-third and one-half of all food produced. Loss and wastage occur at all stages of the food supply chain or value chain. In low-income countries, most loss occurs during production, while in developed countries much food – about 100 kilograms (220 lb) per person per year – is wasted at the consumption stage.

Causes

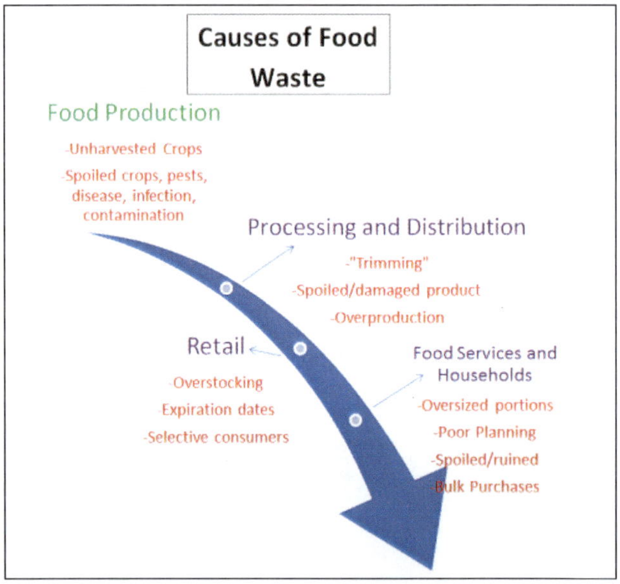

Production

In the US, food waste can occur at most stages of the food industry and in significant amounts. In subsistence agriculture, the amounts of food waste are unknown, but are likely to be insignificant by comparison, due to the limited stages at which waste can occur, and given that food is grown for projected need as opposed to a global marketplace demand. Nevertheless, on-farm losses in storage in developing countries, particularly in African countries, can be high although the exact nature of such losses is much debated.

In the food industry of the United States, the food supply of which is the most diverse and abundant of any country in the world, waste occurs from the beginning of food production chain. From planting, crops can be subjected to pest infestations and severe weather, which cause losses before harvest. Since natural forces (e.g. temperature and precipitation) remain the primary drivers of crop growth, losses from these can be experienced by all forms of outdoor agriculture. On average, farms in the United States lose up to six billion pounds of crops every year because of these unpredictable conditions. The use of machinery in harvesting can cause waste, as harvesters may be unable to discern between ripe and immature crops, or collect only part of a crop. Economic factors, such as regulations and standards for quality and appearance, also cause food waste; farmers often harvest selectively, preferring to leave crops not to standard in the field (where they can be used as fertilizer or animal feed), since they would otherwise be discarded later. This method of removing undesirable produce from harvest collection, distribution sites and grocery stores is called Culling. The USDA defines culling as "the individual removal of genetically undesirable, inferior, weak, diseased, or infested plants from a planting in order to ensure the level of genetic purity or vigor of the crop". However, usually when culling occurs at the production, food processing, retail and consumption stages, it is to remove or dispose of produce with a strange or imperfect appearance rather than produce that is spoiled or unsafe to eat. In urban areas, fruit and nut trees often go unharvested because people either don't realize that the fruit is edible or they fear that it is contaminated, despite research which shows that urban fruit is safe to consume.

Food Processing

Food waste continues in the post-harvest stage, but the amounts of post-harvest loss involved are relatively unknown and difficult to estimate. Regardless, the variety of factors that contribute to food waste, both biological/environmental and socio-economical, would limit the usefulness and reliability of general figures. In storage, considerable quantitative losses can be attributed to pests and micro-organisms. This is a particular problem for countries that experience a combination of heat (around 30 °C) and ambient humidity (between 70 and 90 per cent), as such conditions encourage the reproduction of insect pests and micro-organisms. Losses in the nutritional value, caloric value and edibility of crops, by extremes of temperature, humidity or the action of micro-organisms, also account for food waste. Further losses are generated in the handling of food and by shrinkage in weight or volume.

Some of the food waste produced by processing can be difficult to reduce without affecting the quality of the finished product. Food safety regulations are able to claim foods which contradict standards before they reach markets. Although this can conflict with efforts to reuse food waste (such as in animal feed), safety regulations are in place to ensure the health of the consumer; they are vitally important, especially in the processing of foodstuffs of animal origin (e.g. meat and dairy products), as contaminated products from these sources can lead to and are associated with microbiological and chemical hazards.

Retail

Packaging protects food from damage during its transportation from farms and factories via warehouses to retailing, as well as preserving its freshness upon arrival. Although it avoids considerable food waste, packaging can compromise efforts to reduce food waste in other ways, such as by contaminating waste that could be used for animal feedstocks.

In 2013 the non-profit Natural Resources Defense Council (NRDC) performed research that they state suggests that the leading cause of food waste in America is due to uncertainty over food expiration dates, such as confusion in deciphering best before, sell-by or use-by dates. Joined by Harvard Law, the NRDC produced a study called *The Dating Game: How Confusing Food Date Labels Leads to Food Waste in America*. This United States-based study looked at the intertwining laws which lead labeling to end up unclear and erratic. This uncertainty leads to consumers to toss food, most often because they think the food may be unsafe or misunderstand the labeling on the food completely. Lack of regulation on labeling can result in large quantities of food being removed from the market overall.

Retail stores throw away large quantities of food. Usually, this consists of items that have reached either their best before, sell-by or use-by dates. Food that has passed the best before, and sell-by date, and even some food that passed the use-by date is still edible at the time of disposal, but stores have widely varying policies to handle the excess food. Some stores put effort into preventing access to poor or homeless people, while others work with charitable organizations to distribute food. Retailers also contribute to waste as a result of their contractual arrangements with suppliers. Failure to supply agreed quantities renders farmers or processors liable to have their contracts cancelled. As a consequence, they plan to produce more than actually required to meet the contract, to have a margin of error. Surplus production is often simply disposed of.

Discarded bagels.

Retailers usually have strict cosmetic standards for produce, and if fruits or vegetables are misshapen or superficially bruised, they are often not put on the shelf. In the United States, an estimated six billion pounds of produce is wasted each year because of its appearance. The USDA publishes guidelines used as a baseline assessment by produce distributors, grocery stores, restaurants and other consumers in order to rate the quality of food. These guidelines and how they rate are readily available on their website. For example, apples get graded by their size, color, wax residue, firmness, and skin appearance. If an apples rank highly in these categories and show close to no superficial defects, they are rated as "U.S. Extra Fancy" or "U.S. Fancy", these are the typical ratings sought out by grocery stores when purchasing their produce. Any apples with suboptimal levels of appearance are ranked as either "U.S. Number 1" or "Utility" and are not normally purchased for retail, as recommended by produce marketing sources, despite being safe and edible.

The fish industry also contributes to the annual amount of food waste because of cosmetic standards that the fish are held up to. Nearly "2.3 million tonnes of fish (are) discarded in the North Atlantic and the North Sea each year." Approximately 40 to 60 percent of "all fish caught in Europe is discarded – either because they are the wrong size or species." Addressing this, there are many campaigns focused on raising retailer and consumer awareness about food that fails to meet certain standards for appearance.

Consumption

Consumers are directly and indirectly responsible for wasting a lot of food, which could for a large part be avoided if they were willing to accept suboptimal food (SOF) that deviates in sensory characteristics (odd shapes, discolourations) or has a best-before date that is approaching or has passed, but is still perfectly fine to eat. Cosus (Consumers in a Sustainable food supply chain) is a SUSFOOD ERA-net research project under the topic 'Understanding consumer behaviour to encourage a (more) sustainable food choice'.

Impact on the Environment

According to the Food and Agriculture Organization (FAO), food waste is responsible for 8 percent of global man-made greenhouse gas emissions. The FAO concludes that nearly 30 percent of all available agricultural land in the world - 1.4 billion hectares - is used for produced but uneaten food. The global blue water footprint of food waste is 250 km^3, that is the amount of water that flows annually through the Volga or 3 times Lake Geneva.

Reduction

Reduction in Food Waste.

Response to the problem of food waste at all social levels has varied hugely, including campaigns from advisory and environmental groups, and concentrated media attention on the subject.

Consumer Marketing

One way of dealing with food waste is to reduce its creation. Consumers can reduce spoilage by planning their food shopping, avoiding potentially wasteful spontaneous purchases, and storing foods properly (and also preventing a too large buildup of perishable stock). Widespread educational campaigns have been shown to be an effective way to reduce food waste. A British campaign called "Love Food, Hate Waste" has raised awareness about preventative measures to address food waste for consumers. Through advertisements, information on food storage and preparation and in-store education, the UK observed a 21% decrease in avoidable household food waste over the course of 5 years. Another potential solution is for "smart packaging" which would indicate when food is spoiled more precisely than expiration dates currently do, for example with temperature-sensitive ink, plastic that changes color when exposed to oxygen, or gels that change color with time.

An initiative in Curitiba, Brazil called Cambio Verde allows farmers to provide surplus produce (produce they would otherwise discard due to too low prices) to people that bring glass and metal to recycling facilities (to encourage further waste reduction). In Europe, the Food Surplus Entrepreneurs Network (FSE Network), coordinates a network of social businesses and nonprofit initiatives with the goal to spread best practices to increase the use of surplus food and reduction of food waste.

Collection

In areas where the waste collection is a public function, food waste is usually managed by the same governmental organization as other waste collection. Most food waste is combined with general waste at the source. Separate collections, also known as source-separated organics, have the advantage that food wastes can be disposed of in ways not applicable to other wastes. In the United States, companies find higher and better uses for large commercial generators of food and beverage waste.

From the end of the 19th century through the middle of the 20th century, many municipalities collected food waste (called "garbage" as opposed to "trash") separately. This was typically disinfected by steaming and fed to pigs, either on private farms or in municipal piggeries.

Separate curbside collection of food waste is now being revived in some areas. To keep collection costs down and raise the rate of food waste segregation, some local authorities, especially in Europe, have introduced "alternate weekly collections" of biodegradable waste (including, e.g., garden waste), which enable a wider range of recyclable materials to be collected at reasonable cost, and improve their collection rates. However, they result in a two-week wait before the waste will be collected. The criticism is that particularly during hot weather, food waste rots and stinks, and attracts vermin. Waste container design is therefore essential to making such operations feasible. Curbside collection of food waste is also done in the U.S., some ways by combining food scraps and yard waste together. Several states in the U.S. have introduced a yard waste ban, not accepting leaves, brush, trimmings, etc. in landfills. Collection of food scraps and yard waste combined is then recycled and composted for reuse.

Disposal

As alternatives to landfill, food waste can be composted to produce soil and fertilizer, fed to animals, or used to produce energy or fuel.

Landfills and Greenhouse Gases

Dumping food waste in a landfill causes odour as it decomposes, attracts flies and vermin, and has the potential to add biological oxygen demand (BOD) to the leachate. The European Union Landfill Directive and Waste Regulations, like regulations in other countries, enjoin diverting organic wastes away from landfill disposal for these reasons. Starting in 2015, organic waste from New York City restaurants will be banned from landfills.

In countries such as the United States and the United Kingdom, food scraps constitute around 19% of the waste dumped in landfills, where it ends up rotting and producing methane, a greenhouse gas.

Methane, or CH_4, is the second most prevalent greenhouse gas that is released into the air, also produced by landfills in the U.S. Although methane spends less time in the atmosphere (12 years) than CO_2, it's more efficient at trapping radiation. It is 25 times greater to impact climate change than CO_2 in a 100-year period. Humans accounts over 60% of methane emissions globally.

Animal Feed

Large quantities of fish, meat, dairy and grain are discarded at a global scale annually, when they can be used for things other than human consumption. The feeding of food scraps to domesticated animals is, historically, the most common way of dealing with household food waste. The animals turn roughly two thirds of their ingested food into gas or fecal waste, while the last third is digested and repurposed as meat or dairy products. There are also different ways of growing produce and feeding livestock that could ultimately reduce waste.

Bread and other cereal products discarded from the human food chain could be used to feed chickens. Chickens have traditionally been given mixtures of waste grains and milling by-products in a mixture called chicken scratch. As well, giving table scraps to backyard chickens is a large part of that movement's claim to sustainability, though not all backyard chicken growers recommend it. Ruminants and pigs have also been fed bakery waste for a long time.

Certain food waste (such as flesh) can also be used as feed in maggot farming. The maggots can then be fed to other animals.

Composting

Inevitable waste: peels of potato, onion, lemon, tangerine, banana, kiwi, egg.

Food waste can be biodegraded by composting, and reused to fertilize soil. Composting is the aerobic process completed by microorganisms in which the bacteria break down the food waste into simpler organic materials that can then be used in soil. By redistributing nutrients and high microbial populations, compost reduces water runoff and soil erosion by enhancing rainfall penetration, which has been shown to reduce the loss of sediment, nutrients, and pesticide losses to streams by 75–95%.

Traditional composting uses microbes to perform the decomposition, a process that is most efficient using low, mid, and high temperature microbes (psychrophilic, mesophilic, and thermophilic – respectively). The high temperatures required by the thermophilic microorganisms are hot enough to kill pathogens, making the product of this traditional composting satisfactory for use in soil according to the United States Environmental Protection Agency's (EPA) standards. The traditional decomposition process requires a long length of time and additional energy expended to turn the material frequently and maintain the aerobic process. Composting by thermophilic microbes can lead to nutrient loss and the compost product is heterogeneous, with the potential for higher levels of contaminants which can be harmful if used in agriculture. An alternate method of composting is vermicomposting.

Vermicomposting is the practise of feeding scraps to worms who produce fertilized soil as a by-product. The process of composting using earth worms is completed in a short duration of time and requires no additional energy to turn and maintain aerobic processes, as these actions are already performed by the worms. In order to keep the worms alive, the environment has to be kept below 35 °Celsius, therein making this compost ineligible for use in agriculture according to the EPA standards. The product of vermicomposting is homogenous and generally contains lower levels of contaminants than traditional composting. Some look to integrate vermicomposting and traditional composting in an effort to maximize efficiency while producing a high quality organic product that can be used in agriculture.

Composting food waste leads to a decrease in the quantity of greenhouse gases released into the atmosphere. In landfills, organic food waste decomposes anaerobically, producing methane gas that is emitted into the atmosphere. When this biodegradable waste is composted, it decomposes aerobically and does not produce methane, but instead produces organic compost that can then be utilized in agriculture. Recently, the city of New York has begun to require that restaurants and food-producing companies begin to compost their leftover food. Another instance of composting progress is a Wisconsin-based company called WasteCap, who is dedicated towards aiding local communities create composting plans.

Municipal Food Waste (MFW) can be composted to create this product of organic fertilizer, and many municipalities choose to do this citing environmental protection and economic efficiency as reasoning. Transporting and dumping waste in landfills requires both money and room in the landfills that have very limited available space. One municipality who chose to regulate MFW is San Francisco, who requires citizens to separate compost from trash on their own, instituting fines for non-compliance at $100 for individual homes and $500 for businesses. The city's economic reasoning for this controversial mandate is supported by their estimate that one business can save up to $30000 annually on garbage disposal costs with the implementation of the required composting. Composting is an economical and environmentally conscious step many homeowners could take to reduce their impact on landfill waste. Instead of food scraps and spoiled food taking up space in trashcans or stinking up the kitchen before the bag is full, it could be put outside and broken down by worms and added to garden beds.

Anaerobic Digestion

Anaerobic digestion produces both useful gaseous products and a solid fibrous "compostable" material. Anaerobic digestion plants can provide energy from waste by burning the methane created from food and other organic wastes to generate electricity, defraying the plants' costs and reducing greenhouse gas emissions.The United States Environmental Protection Agency states that the use of anaerobic composting allows for large amounts of food waste to avoid the landfills. Instead of producing these greenhouse gasses into the environment from being in a landfill, the gasses can alternatively be harnessed in these facilities for reuse.

Since this process of composting produces high volumes of biogas, there are potential safety issues such as explosion and poisoning. These interactions require proper maintenance and personal protective equipment is utilized. Certain U.S. states, such as Oregon, have implemented the requirement for permits on such facilities, based on the potential danger to the population and surrounding environment.

Food waste coming through the sanitary sewers from garbage disposal units is treated along with other sewage and contributes to sludge.

Commercial Liquid Food Waste

Commercially, food waste in the form of wastewater coming from commercial kitchens' sinks, dishwashers and floor drains is collected in holding tanks called grease interceptors to minimize flow to the sewer system. This often foul-smelling waste contains both organic and inorganic waste (chemical cleaners, etc.) and may also contain hazardous hydrogen sulfide gases. It is referred to as fats, oils, and grease (FOG) waste or more commonly "brown grease" (versus "yellow grease", which is fryer oil that is easily collected and processed into biodiesel) and is an overwhelming problem, especially in the US, for the aging sewer systems. Per the US EPA, sanitary sewer overflows also occur due to the improper discharge of FOGs to the collection system. Overflows discharge 3 billion US gallons (11,000,000 m^3) – 10 billion US gallons (38,000,000 m^3) of untreated wastewater annually into local waterways, and up to 3,700 illnesses annually are due to exposure to contamination from sanitary sewer overflows into recreational waters.

Agricultural Food Waste

Nearly all global produce, eaten or disposed of is grown using irrigated water. Irrigated water represents the largest sector of water withdraws worldwide, with as much as 90% of total water withdraws being allocated towards agricultural usage. Food which goes uneaten can account for vast quantities of water waste, with food waste being the largest area the average US citizen contributes to water waste.

BIODEGRADABLE PLASTIC

Biodegradable plastics are plastics that can be decomposed by the action of living organisms, usually bacteria.

Two basic classes of biodegradable plastics exist: Bioplastics, whose components are derived from renewable raw materials, and plastics made from petrochemicals containing biodegradable additives which enhance biodegradation.

Examples:

- While aromatic polyesters are almost totally resistant to microbial attack, most aliphatic polyesters are biodegradable due to their potentially hydrolysable ester bonds:
    - Naturally Produced: Polyhydroxyalkanoates (PHAs) like the poly-3-hydroxybutyrate (PHB), polyhydroxyvalerate (PHV) and polyhydroxyhexanoate (PHH);
    - Renewable Resource: Polylactic acid (PLA);
    - Synthetic: Polybutylene succinate (PBS), polycaprolactone (PCL).

- Polyanhydrides,
- Polyvinyl alcohol,
- Most of the starch derivatives,
- Cellulose esters like cellulose acetate and nitrocellulose and their derivatives (such as celluloid, cellophane and rayon),
- Polyethylene terephthalate,
- Enhanced biodegradable plastic with additives.

Development of biodegradable containers.

Home Compostable Plastics

No international standard has been established to define home-compostable plastics, but national standards have been created in Australia (AS 5810 "biodegradable plastics suitable for home composting") and in France (NF T 51-800 "Specifications for plastics suitable for home composting"). The French standard is based on the "OK compost home certification scheme", developed by Belgian certifier TÜV Austria Belgium. The following are examples of plastics that have conformed to an established national standard for home compostability:

- Ecopond Flex 162 resin, maximum thickness 65 micron.
- BWC BF 90A resin, maximum thickness 81 micron.
- BioPBS FD92 resin, maximum thickness 85 micron.
- Torise TRBF90 resin, maximum thickness 43 micron.
- HCPT-1 triple laminate, maximum thickness 119 micron.
- HCFD-2 duplex laminate, maximum thickness 69 micron.

Many people confuse "biodegradable" with "compostable". "Biodegradable" broadly means that an object can be biologically broken down, while "compostable" typically specifies that such a process will result in compost, or humus. Many plastic manufacturers throughout Canada and the US have released products indicated as being compostable. The waste management infrastructure currently recycles regular plastic waste, incinerates it, or places it in a landfill. Mixing biodegradable plastics into the regular waste infrastructure poses some dangers to the environment. However this claim is debatable, if the manufacturer was minimally conforming to the now-withdrawn American Society for Testing and Materials (ASTM) standard definition of the word, as it applies to plastics:

> "That which is capable of undergoing biological decomposition in a compost site such that the material is not visually distinguishable and breaks down into carbon dioxide, water, inorganic compounds and biomass at a rate consistent with known compostable materials."

There is a major discrepancy between this definition and what one would expect from a backyard composting operation. With the inclusion of "inorganic compounds", the above definition allows that the end product might not be humus, an organic substance. The only criterion the ASTM standard definition *did* outline is that a compostable plastic has to become "not visually distinguishable" at the same rate as something that has already been established as being compostable under the traditional definition.

Withdrawal of ASTM D 6002

In January 2011, the ASTM withdrew standard ASTM D 6002, which many plastic manufacturers had been referencing to attain credibility in labelling their products as compostable. The withdrawn description was as follows:

> "This guide covered suggested criteria, procedures, and a general approach to establish the compostability of environmentally degradable plastics."

Advantages and Disadvantages

Under proper conditions, some biodegradable plastics can degrade to the point where microorganisms can completely metabolise them to carbon dioxide (and water). For example, starch-based bioplastics produced from sustainable farming methods could be almost carbon neutral.

There are allegations that biodegradable plastic bags may release metals, and may require a great deal of time to degrade in certain circumstances and that OBD (oxo-biodegradable) plastics may produce tiny fragments of plastic that do not continue to degrade at any appreciable rate regardless of the environment. The response of the Oxo-biodegradable Plastics Association is that OBD plastics do not contain metals. They contain salts of metals, which are not prohibited by legislation and are in fact necessary as trace-elements in the human diet. Oxo-biodegradation of polymer material has been studied in depth at the Technical Research Institute of Sweden and the Swedish University of Agricultural Sciences. A peer-reviewed report of the work shows 91% biodegradation in a soil environment within 24 months, when tested in accordance with ISO 17556.

Environmental Benefits

There is much debate about the total carbon, fossil fuel and water usage in manufacturing bioplastics from natural materials and whether they are a negative impact to human food supply. To make 1 kg (2.2 lb) of polylactic acid, the most common commercially available compostable plastic, 2.65 kg (5.8 lb) of corn is required. Since 270 million tonnes of plastic are made every year, replacing conventional plastic with corn-derived polylactic acid would remove 715.5 million tonnes from the world's food supply, at a time when global warming is reducing tropical farm productivity.

"Although U.S. corn is a highly productive crop, with typical yields between 140 and 160 bushels per acre, the resulting delivery of food by the corn system is far lower. Today's corn crop is mainly used for biofuels (roughly 40 percent of U.S. corn is used for ethanol) and as animal feed (roughly 36 percent of U.S. corn, plus distillers grains left over from ethanol production, is fed to cattle, pigs and chickens). Much of the rest is exported. Only a tiny fraction of the national corn crop is directly used for food for Americans, much of that for high fructose corn syrup."

Traditional plastics made from non-renewable fossil fuels lock up much of the carbon in the plastic, as opposed to being burned in the processing of the plastic. The carbon is permanently trapped inside the plastic lattice, and is rarely recycled, if one neglects to include the diesel, pesticides, and fertilizers used to grow the food turned into plastic.

There is concern that another greenhouse gas, methane, might be released when any biodegradable material, including truly biodegradable plastics, degrades in an anaerobic landfill environment. Methane production from 594 managed landfill environments is captured and used for energy;some landfills burn this off through a process called flaring to reduce the release of methane into the environment. In the US, most landfilled materials today go into landfills where they capture the methane biogas for use in clean, inexpensive energy. Incinerating non-biodegradable plastics will release carbon dioxide as well. Disposing of non-biodegradable plastics made from natural materials in anaerobic (landfill) environments will result in the plastic lasting for hundreds of years.

Bacteria have developed the ability to degrade plastics. While not a solution to the disposal problem, it is likely that bacteria have developed the ability to consume hydrocarbons. In 2008, a 16-year-old boy reportedly isolated two plastic-consuming bacteria.

Environmental Concerns and Benefits

According to a 2010 report of the United States Environmental Protection Agency (EPA) the US had 31 million tons of plastic waste, representing 12.4% of all municipal solid waste. Of that, 2.55 million tons were recovered. This 8.2% recovery was much less than the 34.1% overall recovery percentage for municipal solid waste.

Much of the reason for disappointing plastics recycling goals is that conventional plastics are often commingled with organic wastes (food scraps, wet paper, and liquids), leading to accumulation of waste in landfills and natural habitats. Current usage also makes it difficult and impractical to recycle the underlying polymer without expensive cleaning and sanitizing procedures.

On the other hand, composting of these mixed organics (food scraps, yard trimmings, and wet, non-recyclable paper) is a potential strategy for recovering large quantities of waste and dramatically increasing community recycling goals. As of 2015, food scraps and wet, non-recyclable paper respectively comprise 39.6 million and 67.9 million tons of municipal solid waste. Biodegradable plastics can replace the non-degradable plastics in these waste streams, making municipal composting a significant tool to divert large amounts of otherwise nonrecoverable waste from landfills.

Compostable plastics combine the utility of plastics (lightweight, resistance, relative low cost) with the ability to completely and fully compost in an industrial compost facility. Rather than worrying about recycling a relatively small quantity of commingled plastics, proponents argue that certified biodegradable plastics can be readily commingled with other organic wastes, thereby enabling composting of a much larger portion of nonrecoverable solid waste. Commercial composting for all mixed organics then becomes commercially viable and economically sustainable. More municipalities can divert significant quantities of waste from overburdened landfills since the entire waste stream is now biodegradable and therefore easier to process. This move away from the use of landfills may help alleviate the issue of plastic pollution.

The use of biodegradable plastics, therefore, is seen as enabling the complete recovery of large quantities of municipal solid waste (via aerobic composting and feedstocks) that have heretofore been unrecoverable by other means except land filling or incineration.

Energy Costs for Production

Various researchers have undertaken extensive life cycle assessments of biodegradable polymers to determine whether these materials are more energy efficient than polymers made by conventional fossil fuel-based means. Research done by Gerngross, *et al.* estimates that the fossil fuel energy required to produce a kilogram of polyhydroxyalkanoate (PHA) is 50.4 MJ/kg, which coincides with another estimate by Akiyama, *et al.*, who estimate a value between 50-59 MJ/kg. This information does not take into account the feedstock energy, which can be obtained from non-fossil fuel based methods. Polylactide (PLA) was estimated to have a fossil fuel energy cost of 54-56.7 from two sources, but recent developments in the commercial production of PLA by NatureWorks has eliminated some dependence of fossil fuel-based energy by supplanting it with wind power and biomass-driven strategies. They report making a kilogram of PLA with only 27.2 MJ of fossil fuel-based energy and anticipate that this number will drop to 16.6 MJ/kg in their next generation plants. In contrast, polypropylene and high-density polyethylene require 85.9 and 73.7 MJ/kg, respectively, but these values include the embedded energy of the feedstock because it is based on fossil fuel.

Gerngross reports a 2.65 kg total fossil fuel energy equivalent (FFE) required to produce a single kilogram of PHA, while polyethylene only requires 2.2 kg FFE. Gerngross assesses that the decision to proceed forward with any biodegradable polymer alternative will need to take into account the priorities of society with regard to energy, environment, and economic cost.

Furthermore, it is important to realize the youth of alternative technologies. Technology to produce PHA, for instance, is still in development today, and energy consumption can be further reduced by eliminating the fermentation step, or by utilizing food waste as feedstock. The use of alternative crops other than corn, such as sugar cane from Brazil, are expected to lower energy requirements. For instance, manufacturing of PHAs by fermentation in Brazil enjoys a favorable energy consumption scheme where bagasse is used as source of renewable energy.

Many biodegradable polymers that come from renewable resources (i.e. starch-based, PHA, PLA) also compete with food production, as the primary feedstock is currently corn. For the US to meet its current output of plastics production with BPs, it would require 1.62 square meters per kilogram produced.

Lignin-based Polymer Composites

Lignin-based polymer composites are bio-renewable natural aromatic polymers with biodegradable properties. Lignin is found as a byproduct of polysaccharide extraction from plant material through the production of paper, ethanol, and more. It is high in abundance with reports showing that 50 million tons are being created by chemical pulp industries each year. Lignin is useful due to its low weight material and the fact that it is more environmentally friendly than other alternatives. Lignin is neutral to CO_2 release during the biodegradation process. Other biodegradable plastic processes such as polyethylene terephthalate (PET) have been found to release CO2 and water as waste products produced by the degrading microorganisms.

Lignin contains comparable chemical properties in comparison to current plastic chemicals, which includes reactive functional groups, the ability to form into films, high carbon percentage, and it shows versatility in relation to various chemical mixtures used with plastics. Lignin is also stable, and contains aromatic rings. It is both elastic and viscous yet flows smoothly in the liquid phase. Most importantly lignin can improve on the current standards of plastics because it is antimicrobial in nature. It is being produced at such great quantities and is readily available for use as an emerging environmentally friendly polymer.

Biodegradable Conducting Polymers in the Medical Field

Biodegradable Conducting Polymers (CPs) are a polymeric material designed for applications within the human body. Important properties of this material are its electrical conductivity comparable to traditional conductors and its biodegradability. The medical applications of biodegradable CPs are attractive to medical specialties such as tissue engineering and regenerative medicine. In tissue engineering, the key focus is on providing damaged organs with physicochemical cues to damaged organs for repair. This is achieved through use of nanocomposite scaffolding. Regenerative medicine applications are designed to regenerate cells along with improving the repair process of the body. The use of biodegradable CPs can also be implemented into biomedical imaging along with implants, and more.

The design of biodegradable CPs began with the blending of biodegradable polymers including polylactides, polycaprolactone, and polyurethanes. This design triggered innovation into what is being engineered as of the year 2019. The current biodegradable CPs is applicable for use in the biomedical field. The compositional architecture of current biodegradable CPs includes the conductivity properties of oligomer-based biodegradable polymers implemented into compositions of linear, starshaped, or hyperbranched formations. Another implementation to enhance the biodegradable architecture of the CPs is by use of monomers and conjugated links that are degradable.

The biodegradable polymers used in biomedical applications typically consist of hydrolyzable esters and hydrazones. These molecules, upon external stimulation, go on to be cleaved and broken down. The cleaving activation process can be achieved through use of an acidic environment, increasing the temperature, or by use of enzymes. Three categories of biodegradable CP composites have been established in relation their chemistry makeup. The first category includes partially biodegradable CP blends of conductive and biodegradable polymeric materials. The second category includes conducting oligomers of biodegradable CPs. The third category is that of modified and degradable monimer units along with use of degradable conjugated links for use in biodegradable CPs polymers.

HUMAN WASTE

Human waste (or human excreta) refers to the waste products of the human digestive system and the human metabolism, namely faeces and urine. As part of a sanitation system that is in place, human waste is collected, transported, treated and disposed of or reused by one method or another, depending on the type of toilet being used, ability by the users to pay for services and other factors.

The sanitation systems in place differ vastly across the world, with many people in developing countries having to resort to open defecation where human waste is deposited in the environment, for lack of other options. People in developed countries tend to use flush toilets where the human waste is mixed with water and transported to sewage treatment plants.

Children's excreta can be disposed of in diapers and mixed with municipal solid waste. Diapers are also sometimes dumped directly into the environment, leading to public health risks.

The term "human waste" is used in the general media to mean several things, such as sewage, sewage sludge, blackwater - in fact anything that may contain some human faeces. In the stricter sense of the term, human waste is in fact human excreta, i.e. urine and faeces, with or without water being mixed in. For example, dry toilets collect human waste without the addition of water.

Health Aspects

Human waste is considered a biowaste, as it is a vector for both viral and bacterial diseases. It can be a serious health hazard if it gets into sources of drinking water. The World Health Organization (WHO) reports that nearly 2.2 million people die annually from diseases caused by contaminated water. A major accomplishment of human civilization has been the reduction of disease transmission via human waste through the practice of hygiene and sanitation, which can employ a variety of different technologies.

Environmental Aspects

Even high-mountains are not free from human waste. Each year, millions of mountaineers visit high-mountain areas. They generate tonnes of faeces and urine annually which cause environmental pollution. Human faeces pose a greater threat to the mountain environment than uncontrolled deposit of urine, due to the higher pathogen content of faeces.

Methods of Processing

Methods of processing depend on the type of human waste:

- Sewage is treated via sewage treatment.
- Sewage sludge is treated by sewage sludge treatment.
- Fecal matter from dry toilets may undergo composting.
- Fecal sludge from pit latrines is treated and managed with an approach called fecal sludge management.

The amount of water mixed with human waste can be reduced by the use of waterless urinals and composting toilets and by recycling greywater. The most common method of human waste treatment in rural areas where municipal sewage systems are unavailable is the use of septic tank systems. In remote rural places without sewage or septic systems, small populations allow for the continued use of honey buckets and sewage lagoons without the threat of disease presented by places with denser populations. Bucket toilets are used by rural villages in Alaska where, due to permafrost, conventional waste treatment systems cannot be utilized.

Uses

Human waste in the form of wastewater (sewage) is used to irrigate and fertilize fields in many parts of the developing world where fresh water is unavailable. There is great potential for wastewater agriculture to produce more food for consumers in urban areas, as long as there is sufficient education about the dangers of eating such food uncooked.

SEWAGE

Sewage (or domestic wastewater or municipal wastewater) is a type of wastewater that is produced by a community of people. It is characterized by volume or rate of flow, physical condition, chemical and toxic constituents, and its bacteriologic status (which organisms it contains and in what quantities). It consists mostly of greywater (from sinks, tubs, showers, dishwashers, and clothes washers), blackwater (the water used to flush toilets, combined with the human waste that it flushes away); soaps and detergents; and toilet paper (less so in regions where bidets are widely used instead of paper).

Sewage usually travels from a building's plumbing either into a sewer, which will carry it elsewhere, or into an onsite sewage facility (of which there are many kinds). Whether it is combined with surface runoff in the sewer depends on the sewer design (sanitary sewer or combined sewer). The reality is that most wastewater produced globally remains untreated causing widespread water pollution, especially in low-income countries: A global estimate by UNDP and UN-Habitat is that 90% of all wastewater generated is released into the environment untreated. In many developing countries the bulk of domestic and industrial wastewater is discharged without any treatment or after primary treatment only.

The term sewage is nowadays regarded as an older term and is being more and more replaced by "wastewater". In general American English usage, the terms "sewage" and "sewerage" mean the same thing. In common British usage, and in American technical and professional English usage, "sewerage" refers to the infrastructure that conveys sewage.

Types

Greywater samples before and after treatment in a constructed wetland in Peru.

- The wastewater from residences and institutions, carrying bodily wastes (primarily feces and urine), washing water, food preparation wastes, laundry wastes, and other waste products of normal living, are classed as domestic or sanitary sewage.
- Liquid-carried wastes from stores and service establishments serving the immediate community, termed commercial wastes, are included in the sanitary or domestic sewage category if their characteristics are similar to household flows. Wastes that result from industrial processes such as the production or manufacture of goods are classed as industrial wastewater, not as sewage.
- Surface runoff, also known as storm flow or overland flow, is that portion of precipitation that runs rapidly over the ground surface to a defined channel. Precipitation absorbs gases and particulates from the atmosphere, dissolves and leaches materials from vegetation and soil, suspends matter from the land, washes spills and debris from urban streets and highways, and carries all these pollutants as wastes in its flow to a collection point.

Pollutants

Organic Pollutants and Nutrients

Sewage is a complex mixture of chemicals, with many distinctive chemical characteristics. These include high concentrations of ammonium, nitrate, nitrogen, phosphorus, high conductivity (due to high dissolved solids), high alkalinity, with pH typically ranging between 7 and 8. The organic matter of sewage is measured by determining its biological oxygen demand (BOD) or the chemical oxygen demand (COD).

Pathogens

Sewage contains human feces, and therefore often contains pathogens of one of the four types:

- Bacteria (for example Salmonella, Shigella, Campylobacter, Vibrio cholerae),
- Viruses (for example hepatitis A, rotavirus, enteroviruses),
- Protozoa (for example Entamoeba histolytica, Giardia lamblia, Cryptosporidium parvum),
- Parasites such as helminths and their eggs (e.g. ascaris (roundworm), ancylostoma (hookworm), trichuris (whipworm)).

Sewage can be monitored for both disease-causing and benign organisms with a variety of techniques. Traditional techniques involve filtering, staining, and examining samples under a microscope. Much more sensitive and specific testing can be accomplished with DNA sequencing, such as when looking for rare organisms, attempting eradication, testing specifically for drug-resistant strains, or discovering new species. Sequencing DNA from an environmental sample is known as metagenomics.

Micro-pollutants

Sewage also contains environmental persistent pharmaceutical pollutants. Trihalomethanes can also be present as a result of past disinfection.

Sewage has also been analyzed to determine relative rates of use of prescription and illegal drugs among municipal populations.

Health and Environmental Aspects

All categories of sewage are likely to carry pathogenic organisms that can transmit disease to humans and animals. Sewage also contains organic matter that can cause odor and attract flies.

Sewage contains nutrients that may cause eutrophication of receiving water bodies; and can lead to ecotoxicity.

Collection

A system of sewer pipes (sewers) collects sewage and takes it for treatment or disposal. The system of sewers is called *sewerage* or *sewerage system* in British English and *sewage system* in American English. Where a main sewerage system has not been provided, sewage may be collected from homes by pipes into septic tanks or cesspits, where it may be treated or collected in vehicles and taken for treatment or disposal. Properly functioning septic tanks require emptying every 2–5 years depending on the load of the system.

A medieval waste pipe in Stockholm Old Town formerly deposited sewage on the street to be flushed away by rain.

Sewage canal of a medieval house as depicted in 1447 *St. Barbara Altarpiece* in the National Museum in Warsaw.

Treatment

Sewage treatment is the process of removing the contaminants from sewage to produce liquid and solid (sludge) suitable for discharge to the environment or for reuse. It is a form of waste management. A septic tank or other on-site wastewater treatment system such as biofilters or constructed wetlands can be used to treat sewage close to where it is created.

Sewage treatment results in sewage sludge which requires sewage sludge treatment before safe disposal or reuse. Under certain circumstances, the treated sewage sludge might be termed "biosolids" and can be used as a fertilizer.

In developed countries sewage collection and treatment is typically subject to local and national regulations and standards.

Disposal

Raw sewage is also disposed of to rivers, streams, and the sea in many parts of the world. Doing so can lead to serious pollution of the receiving water. This is common in developing countries and may still occur in some developed countries, for various reasons – usually related to costs.

Ships at sea are forbidden from discharging their sewage overboard unless three miles or more from shore.

Reuse of Treated or Untreated Sewage

Increasingly, agriculture is using untreated wastewater for irrigation. Cities provide lucrative markets for fresh produce, so are attractive to farmers. Because agriculture has to compete for increasingly scarce water resources with industry and municipal users, there is often no alternative for farmers but to use water polluted with urban waste, including sewage, directly to water their crops. There can be significant health hazards related to using water loaded with pathogens in this way, especially if people eat raw vegetables that have been irrigated with the polluted water.

The International Water Management Institute has worked in India, Pakistan, Vietnam, Ghana, Ethiopia, Mexico and other countries on various projects aimed at assessing and reducing risks of wastewater irrigation. They advocate a 'multiple-barrier' approach to wastewater use, where farmers are encouraged to adopt various risk-reducing behaviours. These include ceasing irrigation a few days before harvesting to allow pathogens to die off in the sunlight, applying water carefully so it does not contaminate leaves likely to be eaten raw, cleaning vegetables with disinfectant or allowing fecal sludge used in farming to dry before being used as a human manure. The World Health Organization has developed guidelines for safe water use.

SEWAGE SLUDGE

Sewage sludge is the residual, semi-solid material that is produced as a by-product during sewage treatment of industrial or municipal wastewater. The term "septage" also refers to sludge from simple wastewater treatment but is connected to simple on-site sanitation systems, such as septic tanks.

When fresh sewage or wastewater enters a primary settling tank, approximately 50% of the suspended solid matter will settle out in an hour and a half. This collection of solids is known as raw sludge or primary solids and is said to be "fresh" before anaerobic processes become active. The sludge will become putrescent in a short time once anaerobic bacteria take over, and must be removed from the sedimentation tank before this happens.

This is accomplished in one of two ways. In an Imhoff tank, fresh sludge is passed through a slot to the lower story or digestion chamber where it is decomposed by anaerobic bacteria, resulting in liquefaction and reduced volume of the sludge. After digesting for an extended period, the result is called "digested" sludge and may be disposed of by drying and then landfilling. More commonly with domestic sewage, the fresh sludge is continuously extracted from the tank mechanically and passed to separate sludge digestion tanks that operate at higher temperatures than the lower story of the Imhoff tank and, as a result, digest much more rapidly and efficiently.

"Biosolids" is a term often used in conjunction with reuse of sewage solids after sewage sludge treatment. Biosolids can be defined as organic wastewater solids that can be reused after stabilization processes such as anaerobic digestion and composting. Opponents of sewage sludge reuse reject this term as a public relations term.

Classes of Treated Sewage Sludge (US)

Class A sludge is typically dried and pasteurized, and is also known as "exceptional" quality.

Class B includes all sludge not classified as Class A. Class B sludge is typically "undigested" and is volatile.

Both classes of sludge may still contain radioactive or pharmaceutical wastes.

Quantities Produced

The amount of sewage sludge produced is proportional to the amount and concentration of wastewater treated, and it also depends on the type of wastewater treatment process used. It can be expressed as kg dry solids per cubic metre of wastewater treated. The total sludge production from a wastewater treatment process is the sum of sludge from primary settling tanks (if they are part of the process configuration) plus excess sludge from the biological treatment step. For example, primary sedimentation produces about 110–170 kg/ML of so-called primary sludge, with a value of 150 kg/ML regarded as being typical for municipal wastewater in the U.S. or Europe. The sludge production is expressed as kg of dry solids produced per ML of wastewater treated; one mega litre (ML) is 10^3 m^3. Of the biological treatment processes, the activated sludge process produces about 70–100 kg/ML of waste activated sludge, and a trickling filter process produces slightly less sludge from the biological part of the process: 60–100 kg/ML. This means that the total sludge production of an activated sludge process that uses primary sedimentation tanks is in the range of 180–270 kg/ML, being the sum of primary sludge and waste activated sludge.

United States municipal wastewater treatment plants in 1997 produced about 7.7 million dry tons of sewage sludge, and about 6.8 million dry tons in 1998 according to EPA estimates. As of 2004, about 60% of all sewage sludge was applied to land as a soil amendment and fertilizer for growing crops.

Production of sewage sludge can be reduced by conversion from flush toilets to dry toilets such urine-diverting dry toilets and composting toilets.

Contaminants

Pathogens

Bacteria in Class A sludge products can actually regrow under certain environmental conditions. Pathogens could easily remain undetected in untreated sewage sludge. Pathogens are not a significant health issue if sewage sludge is properly treated and site-specific management practices are followed.

Micro-pollutants

Micro-pollutants can become concentrated in sewage sludge. Each of these disposal options comes with myriad potential — and in some cases proven — human health and environment impacts.

Sterols and other hormones have been detected.

Heavy Metals

One of the main concerns in the treated sludge is the concentrated metals content (arsenic, cadmium, copper, etc., some of which are also critical plant micronutrients; certain metals are regulated while others are not. Leaching methods can be used to reduce the metal content and meet the regulatory limit.

In 2009 the EPA released the Targeted National Sewage Sludge Study, which reports on the level of metals, chemicals, hormones, and other materials present in a statistical sample of sewage sludges. Some highlights include:

- Silver is present to the degree of 20 mg/kg of sludge, on average, a near economically recoverable level, while some sludges of *exceptionally high quality* have up to 200 milligrams of silver per kilogram of sludge; one outlier demonstrated a silver lode of 800–900 mg per kg of sludge.
- Barium is present at the rate of 500 mg/kg, while manganese is present at the rate of 1 g/kg sludge.
- Lead, arsenic, chromium, and cadmium are estimated by the EPA to be present in detectable quantities in 100% of national sewage sludges in the US, while thallium is only estimated to be present in 94.1% of sludges.

Other Hazardous Substances

Sewage treatment plants receive various forms of hazardous waste from hospitals, nursing homes, industry and households. Low levels of constituents such as PCBs, dioxin, and brominated flame retardants, may remain in treated sludge. There are potentially thousands other components of sludge that remain untested/undetected disposed of from modern society that also end up in sludge (pharmaceuticals, nano particles, etc.) which have been proven to be hazardous to both human and ecological health.

In 2013 in South Carolina PCBs were discovered in very high levels in wastewater sludge. The problem was not discovered until thousands of acres of farm land in South Carolina were contaminated by this hazardous material. SCDHEC issued emergency regulatory order banning all PCB laden sewage sludge from being land applied on farm fields or deposited into landfills in South Carolina.

Also in 2013, after DHEC request, the city of Charlotte decided to stop land applying sewage sludge in South Carolina while authorities investigated the source of PCB contamination. In February 2014, the city of Charlotte admitted PCBs have entered their sewage treatment centers as well.

Contaminants of concern in sewage sludge are plasticizers, PDBEs, and others generated by human activities, including personal care products and medicines. Synthetic fibers from fabrics persist in treated sewage sludge as well as in biosolids-treated soils and may thus serve as an indicator of past biosolids application.

Pollutant Ceiling Concentration

The term "pollutant" is defined as part of the EPA 503 rule. The components of sludge have pollutant limits defined by the EPA. "A Pollutant is an organic substance, an inorganic substance, a combination of organic and inorganic substances, or a pathogenic organism that, after discharge and upon exposure, ingestion, inhalation, or assimilation into an organism either directly from the environment or indirectly by ingestion through the food chain, could, on the basis of information available to the Administrator of EPA, cause death, disease, behavioral abnormalities, cancer, genetic mutations, physiological malfunctions (including malfunction in reproduction), or physical deformations in either organisms or offspring of the organisms." The maximum component pollutant limits by the US EPA are:

Pollutant	Ceiling concentration (milligrams per kilogram)
Cadmium	85
Copper	4300
Lead	840
Mercury	57
Molybdenum	75
Nickel	420
Selenium	100
Zinc	7500

Treatment

Sewage sludge treatment is the process of removing contaminants from wastewater. Sewage sludge is produced from the treatment of wastewater in sewage treatment plants and consists of two basic forms — raw primary sludge and secondary sludge, also known as activated sludge in the case of the activated sludge process.

Sewage sludge is usually treated by one or several of the following treatment steps: lime stabilization, thickening, dewatering, drying, anaerobic digestion or composting. Some treatment processes, such as composting and alkaline stabilization, that involve significant amendments may affect contaminant strength and concentration: depending on the process and the contaminant in question, treatment may decrease or in some cases increase the bioavailability and/or solubility of contaminants.

Following treatment, sewage sludge is either landfilled, incinerated, applied on agricultural land or, in some cases, retailed or given away for free to the general public.

Health Risks

In 2011, the EPA commissioned a study at the United States National Research Council (NRC) to determine the health risks of sludge. In this document the NRC pointed out that many of the dangers of sludge are unknown and unassessed. Additionally "Regulations that limit contact with biosolids do not prevent environmental processes in the conceptual model such as aerosolization or erosion and the death or multiplication of pathogens."

The National Research Council published "Biosolids Applied to Land: Advancing Standards and Practices" in July 2002. The NRC concluded that while there is no documented scientific evidence that sewage sludge regulations have failed to protect public health, there is persistent uncertainty on possible adverse health effects. The NRC noted that further research is needed and made about 60 recommendations for addressing public health concerns, scientific uncertainties, and data gaps in the science underlying the sewage sludge standards. The EPA responded with a commitment to conduct research addressing the NRC recommendations.

Residents living near Class B sludge processing sites may experience asthma or pulmonary distress due to bioaerosols released from sludge fields.

A 2004 survey of 48 individuals near affected sites found that most reported irritation symptoms, about half reported an infection within a month of the application, and about a fourth were affected by *Staphylococcus aureus*, including two deaths. The number of reported S. aureus infections was 25 times as high as in hospitalized patients, a high-risk group. The authors point out that regulations call for protective gear when handling Class B biosolids and that similar protections could be considered for residents in nearby areas given the wind conditions.

Khuder, Milz, Bisesi, Vincent, McNulty, and Czajkowski (as cited by Harrison and McBride of the Cornell Waste Management Institute in *Case for Caution Revisited: Health and Environmental Impacts of Application of Sewage Sludges to Agricultural Land*) conducted a health survey of persons living in close proximity to Class B sludged land. A sample of 437 people exposed to Class B sludge (living within 1-mile (1.6 km) of sludged land) - and using a control group of 176 people not exposed to sludge (not living within 1-mile (1.6 km) of sludged land) reported the following:

> "Results revealed that some reported health-related symptoms were statistically significantly elevated among the exposed residents, including excessive secretion of tears, abdominal bloating, jaundice, skin ulcer, dehydration, weight loss, and general weakness. The frequency of reported occurrence of bronchitis, upper respiratory infection, and giardiasis were also statistically significantly elevated. The findings suggest an increased risk for certain respiratory, gastrointestinal, and other diseases among residents living near farm fields on which the use of biosolids was permitted."

Although correlation does not imply causation, such extensive correlations may lead reasonable people to conclude that precaution is necessary in dealing with sludge and sludged farmlands.

Harrison and Oakes suggest that, in particular, "until investigations are carried out that answer these questions (about the safety of Class B sludge), land application of Class B sludges should be viewed as a practice that subjects neighbors and workers to substantial risk of disease." They further suggest that even Class A treated sludge may have chemical contaminants (including heavy metals, such as lead) or endotoxins present, and a precautionary approach may be justified on this basis, though the vast majority of incidents reported by Lewis, et al. have been correlated with exposure to Class B untreated sludge and not Class A treated sludge.

A 2005 report by the state of North Carolina concluded that "a surveillance program of humans living near application sites should be developed to determine if there are adverse health effects in humans and animals as a result of biosolids application."

Disposal

After treatment, and dependent upon the quality of sludge produced (for example with regards to heavy metal content), sewage sludge is most commonly either disposed of in landfills, dumped in the ocean or applied to land for its fertilizing properties, as pioneered by the product Milorganite.

Landfill

Sewage sludge deposition in landfills can circulate human-virulent species of Cryptosporidium and Guardia pathogens. Sonication and quicklime stabilization are most effective in inactivation of these pathogens; microwave energy disintegration and top-soil stabilization were less effective.

Ocean Dumping

It used to be common practice to dump sewage sludge into the ocean, however, this practice has stopped in many nations due to environmental concerns as well to domestic and international laws and treaties.

Land Application

Biosolids is a term widely used to denote the byproduct of domestic and commercial sewage and wastewater treatment that is to be used in agriculture. National regulations that dictate the practice of land application of treated sewage sludge differ widely and e.g. in the US there are widespread disputes about this practice.

Depending on their level of treatment and resultant pollutant content, biosolids can be used in regulated applications for non-food agriculture, food agriculture, or distribution for unlimited use. Treated biosolids can be produced in cake, granular, pellet, or liquid form and are spread over land before being incorporated into the soil or injected directly into the soil by specialist contractors. Such use was pioneered by the production of Milorganite in 1926.

Use of sewage sludge has shown an increase in level of soil available phosphorus and soil salinity.

The findings of a 20-year field study of air, land, and water in Arizona, concluded that use of biosolids is sustainable and improves the soil and crops. Other studies report that plants uptake large quantities of heavy metals and toxic pollutants that are retained by produce, which is then consumed by humans.

A PhD thesis studying the addition of sludge to neutralize soil acidity concluded that the practice was not recommended if large amounts are used because the sludge produces acids when it oxidizes.

Studies have indicated that pharmaceuticals and personal care products, which often adsorb to sludge during wastewater treatment, can persist in agricultural soils following biosolid application. Some of these chemicals, including potential endocrine disruptor Triclosan, can also travel through the soil column and leach into agricultural tile drainage at detectable levels. Other studies, however, have shown that these chemicals remain adsorbed to surface soil particles, making them more susceptible to surface erosion than infiltration. These studies are also mixed in their findings regarding the persistence of chemicals such as triclosan, triclocarban, and other pharmaceuticals. The impact of this persistence in soils is unknown, but the link to human and land animal health is likely tied to the capacity for plants to absorb and accumulate these chemicals in their consumed tissues. Studies of this kind are in early stages, but evidence of root uptake and translocation to leaves did occur for both triclosan and triclocarban in soybeans. This effect was not present in corn when tested in a different study.

A cautionary approach to land application of biosolids has been advocated by some for regions where soils have lower capacities for toxics absorption or due to the presence of unknowns in sewage biosolids. In 2007 the Northeast Regional Multi-State Research Committee (NEC 1001) issued conservative guidelines tailored to the soils and conditions typical of the northeastern US.

Use of sewage sludge is prohibited for produce to be labeled USDA-certified organic. In 2014 the United States grocery chain Whole Foods banned produce grown in sewage sludge.

Treated sewage sludge has been used in the UK, Europe and China agriculturally for more than 80 years, though there is increasing pressure in some countries to stop the practice of land application due to farm land contamination and public outrage. In the 1990s there was pressure in some European countries to ban the use of sewage sludge as a fertilizer. Switzerland, Sweden, Austria, and others introduced a ban. Since the 1960s there has been cooperative activity with industry to reduce the inputs of persistent substances from factories. This has been very successful and, for example, the content of cadmium in sewage sludge in major European cities is now only 1% of what it was in 1970.

Incineration

Sludge can also be incinerated in sludge incineration plants which comes with its own set of environmental concerns (air pollution, disposal of the ash). Pyrolysis of the sludge to create syngas and potentially biochar is possible, as is combustion of biofuel produced from drying sewage sludge or incineration in a waste-to-energy facility for direct production of electricity and steam for district heating or industrial uses.

Thermal processes can greatly reduce the volume of the sludge, as well as achieve remediation of all or some of the biological concerns. Direct waste-to-energy incineration and complete combustion systems (such as the Gate 5 Energy System) will require multi-step cleaning of the exhaust gas, to ensure no hazardous substances are released. In addition, the ash produced by incineration or incomplete combustion processes (such as fluidized-bed dryers) may be difficult to use without subsequent treatment due to high heavy metal content; solutions to this include leaching of the ashes to remove heavy metals or in the case of ash produced in a complete-combustion process, or with biochar produced from a pyrolytic process, the heavy metals may be fixed in place and the ash material readily usable as a LEEDs preferred additive to concrete or asphalt. Examples of other ways to use dried sewage sludge as an energy resource include the Gate 5 Energy System, an innovative process to power a steam turbine using heat from burning milled and dried sewage sludge, or combining dried sewage sludge with coal in coal-fired power stations. In both cases this allows for production of electricity with less carbon-dioxide emissions than conventional coal-fired power stations.

FECES

Feces, faeces, excrement or dung is a waste product from an animal or human. Feces is an extremely rich source of soil nutrients and soil microorganisms, and treating it as waste discards a valuable resource. However it is also a potent source of human pathogens, and attractive to disease vectors such as flies, making it dangerous to public health. Therefore it must be appropriately treated in most cases and applied according to safety guidelines.

Feces, when used as a resource is referred to as manure. Similarly, human waste, when treated in the same context is referred to as humanure. Humanure must be handled with particular care is it is more likely to carry human pathogens, compared to animal manure.

Uses

Feces can be used for several purposes:

- It can be used as is, as an organic fertilizer in agricultural fields (the feces is then called "manure") Use as is (as night soil) however is known to cause illness, especially among those not accustomed to eating food grown in this way.
- It can be composted to form humus (ie using a composting toilet): this is slightly more environmentally friendly.
- Rather than using it in pure form or as compost, only the phosporus too can be extracted by adding magnesium to form struvite and then collect the struvite (less environmentally friendly).
- It can be used in biodigesters to form biogas and manure (residue).

Note that water-based systems require sewage treatment and use of this is best avoided. Also note that manure can also be allowed to be digested by composting worms (e.g. for feeding fish).

Feces as Fuel

Dung is considered to be one of the best fuels for the traditional mud stove for the following reasons:

- It burns slowly.
- Cooks fast.
- Generates powerful heat compared to other sources of fuel found locally.
- Easy to store.
- Less toxic compared to other sources of fuel.

Problems related to dung as a fuel are:

- There is a scarcity of dung.

- Cattle owners do not permit collection form their fields.
- As dung is being dried there is a risk that it could be stolen.
- It burns faster than wood when it is not properly compressed.

Making Clean Dung Briquettes from Feces

Dung washing is a simple process whereby the dung is washed with water, organic matter, fermented paper, sawdust or other additives mixed into the slurry and then the liquid removed via compression. The benefits of washed compressed dung as a fuel source are three fold:

- The chlorine and the silica present in the dung are water soluble and hence can be removed prior to burning. These components contribute a serious health risk for users when burnt.
- The water-soluble fraction of the dung also constitutes the "agriculturally nutritious" aspects of dung and thus can be retained for fertiliser.
- Removal of the silica content, lowers the ash content which inturn allows for the proper flow of air which overall translates to a reduction in the formation of the harmful carbon monoxide.

Two stage press being used by village women in Rupendehi, Nepal.

A briquette press is a tool that can be used to remove the liquid and bind the slurry into solid briquettes. A two stage press has been designed with a hybrid lever and screw compression that produces briquettes fast and efficiently with the minimum of effort. Initially a lever is depressed to a comfortable operating level, then the lever mechanism is locked in place and a secondary screw system further depresses a plunger and applies the additional force necessary to to remove additional moisture. The steel briquette press can be manufactured locally, is easily maintenaned and simple to use. The Briquette chamber can produce four briquettes simultaneously. Having two briquette chambers and base stands allows for increased speed an efficiency as one chamber can be filled whilst the other is being compressed. Briquettes take on average 6 days to dry in dry temperate conditions.

Safe use of Humanure

It is an important principle in risk management to take several safety measures together, to minimize risk. Although humanure is relatively safe, extra precautions should be taken for the possibility of pathogens surviving the composting process, and for those with weakened immune systems.

The key safety measures involve:

- Processing: ensuring pathogens are killed, through temperature, time and pH.
- Separation: ensuring that the humanure and any remaining pathogens don't touch or enter food.
- Food treatment: Washing, peeling and/or cooking reduce the risk further, if necessary.

For food safety, it is recommended to apply mature humanure to an empty garden bed first, then apply layers of safer soil and mulch.

Another good strategy is to use thick mulch, and only apply where the edible parts of the plant are well above the ground - fruit trees or vines on trellises, for example.

Variations in Treatment Requirements

Treatment and application requirements will be less strict where the soil will not be used for food crops for months or years, and where there is little exposure to humans and human disease vectors. In such cases, appropriate precautions are still important, such as mixing with soil or covering with a later of soil, to speed breakdown. Mulch adds physical protection, and compost tea might be expected to affair the breakdown and suppression of pathogens.

The Humanure Handbook by Joseph Jenkins explains how to compost human feces in such a way as to kill the pathogens. A summary of the method is:

- Use sawdust to bury manure. Sawdust provides carbon and this balances the nitrogen in the faeces. A good carbon-nitrogen balance is needed for composting. The Sawdust also keeps bad smells in and keeps flies away.
- Do not separate urine and feces, as the urine provides moisture the microorganisms need.
- As well as adding the buckets from the toilet to the compost heap, throw on chunky materials like straw. This traps air pockets in the compost heap.
- Optionally, skewer the compost heap with sticks or pipes to provide more aeration.
- Let it sit for a year before using on plants.

Note that most approaches however emphasize separating the urine, as otherwise the nitrogen content becomes too high for proper composting (requires a specific carbon/nitrogen ratio).

ORGANIC WASTE

Organic wastes contain materials which originated from living organisms. There are many types of organic wastes and they can be found in municipal solid waste, industrial solid waste, agricultural waste, and wastewaters. Organic wastes are often disposed of with other wastes in landfills or incinerators, but since they are biodegradable, some organic wastes are suitable for composting and land application.

Organic materials found in municipal solid waste include food, paper, wood, sewage sludge, and yard waste. Because of recent shortages in landfill capacity, the number of municipal composting sites for yard wastes is increasing across the country, as is the number of citizens who compost yard wastes in their backyards. On a more limited basis, some mixed municipal waste composting is also taking place. In these systems, attempts to remove inorganic materials are made prior to composting.

Some of the organic materials in municipal solid waste are separated before disposal for purposes other than composting. For example, paper and cardboard are commonly removed for recycling. Food waste from restaurants and grocery stores is typically disposed of through garbage disposals, therefore, it becomes a component of wastewater and sewage sludge.

A large percentage of sewage sludge is landfilled and incinerated, but it is increasingly being applied to land as a fertilizer. Sewage sludge may be used as an agricultural fertilizer or as an aid in reclaiming land devastated by strip mining, deforestation, and over-application of inorganic fertilizers. It may also be applied to land solely as a means of disposal, without the intention of improving the soil.

The organic fraction of industrial waste covers a wide spectrum including most of the components of municipal organic waste, as well as countless other materials. A few examples of industrial organic wastes are papermill sludge, meat processing waste, brewery wastes, and textile mill fibers. Since a large variety and volume of industrial organic wastes are generated, there is a lot of potential to recycle and compost these materials. Waste managers are continually experimenting with different "recipes" for composting industrial organic wastes into soil conditioners and soil amendments. Some treated industrial wastewaters and sludges contain large amounts of organic materials and they too can be used as soil fertilizers and amendments.

Production of biogas is another use of organic waste. Biogas is used as an alternative energy source in some third world countries. It is produced in digester units by the anaerobic decomposition of organic wastes such as manures and crop residues. Beneficial byproducts of biogas production include sludges that can be used to fertilize and improve soil, and the inactivation of pathogens in the waste. In addition, there is ongoing research on using organic wastes in developing countries: in fish farming; to produce algae for human and animal consumption, fertilizer, and other uses; and to produce aquatic macrophytes for animal feed supplements.

RECYCLING ORGANIC WASTE

Organic Waste – Types, Sources and Uses

There a number of types of organic waste which are commonly discarded. We will look at the types and sources of organic waste and some examples of common uses for this waste.

Domestic or Household Waste

This type of waste is usually made up of food scraps, either cooked or uncooked, and garden waste such as grass cuttings or trimmings from bushes and hedges. Domestic kitchen waste is often mixed with non-organic materials such as plastic packaging, which cannot be composted. It is beneficial if this type of waste can be separated at source – this makes recycling of both types of waste far easier. Domestic or household waste is usually produced in relatively small quantities. In developing countries, there is a much higher organic content in domestic waste. From Figure we can see that up to 60% (or more in some cases) of all municipal waste is organic matter, much higher than the figure for an industrialised country. It is therefore well worth intercepting this supply of useful material where it can be used effectively.

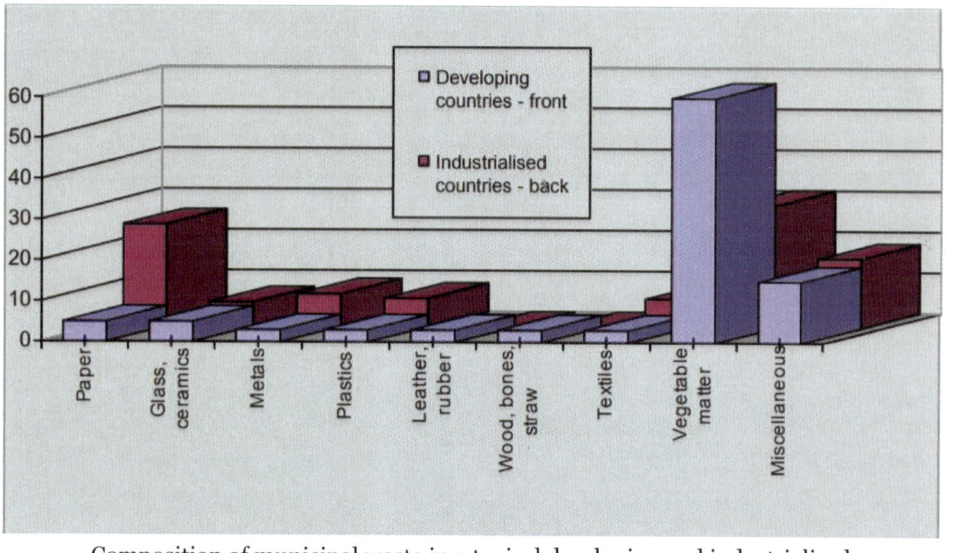

Composition of municipal waste in a typical developing and industrialised country (actual figures vary significantly – this figure is only an example).

Commercially Produced Organic Waste

By this, we mean waste generated at institutional buildings, such as schools, hotels and restaurants. The quantities of waste here are much higher and the potential for use in conjunction with small- scale enterprise is good.

Animal and Human Waste

It is worth mentioning at the start of this topic that there are serious health risks involved with handling sewage. Raw sewage contains bacteria and pathogens that cause serious illness and disease. It should be stressed that health and safety procedures should be followed when dealing with sewage and that people involved with its handling should have a clear understanding of the health risks involved. Raw sewage should never be applied to crops which are for consumption by humans or animals.

- Human faecal residue is produced in large quantities in urban areas and is dealt with in a variety of ways. In the worst cases, little is done to remove or treat the waste and it can present enormous health risks. This is often the case in the slum districts or poor areas of some large cities. Sewage is often dealt with crudely and is pumped into the nearest water body with little or no treatment. There are methods for large-scale treatment and use of sewage as a fertiliser and a source of energy. The most commonly used method is anaerobic digestion to produce biogas and liquid fertiliser. Composting toilets facilitate the conversion of human faecal waste into rich compost.

- Animal residue is rarely wasted. This fertile residue is commonly used as a source of fertiliser, being applied directly to the land, or as a source of energy, either through direct combustion (after drying) or through digestion to produce methane gas.

Agricultural Residue

This is the 'waste' which remains after the processing of crops (e.g. maize stalks, rice husks, foliage, etc.). There are a wide variety of applications for this residue, ranging from simple combustion on an open fire to complex energy production processes that use this waste as a fuel stock. It is not within the scope of this paper to deal with the many and varied uses of agricultural residues.

Methods of Processing Organic Waste

As mentioned in the introduction, there are three main ways in which organic waste can be used:

- Soil Improvement,
- Animal raising,
- To provide a source of energy.

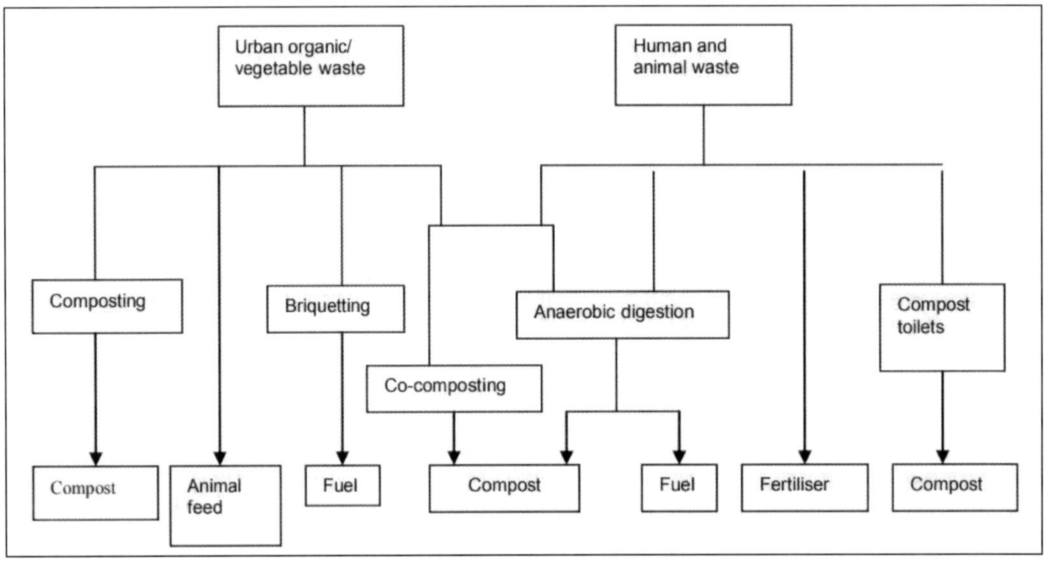

Processes and products from organic waste.

Differing levels of processing are required for achieving the above and in this topic we will take a brief look at just some of the common approaches to using organic waste. Figure shows some of the options in the form of a flow diagram.

References

- Tchobanoglous, george; burton, franklin l.; stensel, h. David (2003). Wastewater engineering : treatment and reuse (4 ed.). Metcalf & eddy. P. 1449. Isbn 978-0071122504
- Williams, i.d.; kelly, j. (2003). "green waste collection and the public's recycling behaviour in the borough of wyre, england". Resources, conservation and recycling. 38 (2): 139–159. Doi:10.1016/s0921-3449(02)00106-4. Issn 0921-3449
- What-is-green-waste, skipthetip.com, Retrieved 25 August, 2019
- Belyaeva, o.n.; haynes, r.j. (2009). "chemical, microbial and physical properties of manufactured soils produced by co-composting municipal green waste with coal fly ash". Bioresource technology. 100 (21): 5203–5209. Doi:10.1016/j.biortech.2009.05.032. Issn 0960-8524. Pmid 19539464
- Feces: appropedia.org, Retrieved 4 June, 2019
- Noble, r.; coventry, e. (2005). "suppression of soil-borne plant diseases with composts: a review". Biocontrol science and technology. 15 (1): 3–20. Doi:10.1080/09583150400015904. Issn 0958-3157
- Guide to the facts and figures report about materials, waste and recycling". Epa. 2017-09-07. Retrieved 8 sep 2018
- Avérous, luc; pollet, eric (2014), "nanobiocomposites based on plasticized starch", starch polymers, elsevier, pp. 211–239, doi:10.1016/b978-0-444-53730-0.00028-2, isbn 9780444537300
- Organic-waste, encyclopedias-almanacs-transcripts-and-maps, environment: encyclopedia.com, Retrieved 12 April, 2019
- Wastewater engineering: treatment and reuse (4th ed.). Metcalf & eddy, inc., mcgraw hill, usa. 2003. P. 1807. Isbn 0-07-112250-8

3

Types of Biodegradable Waste Management

Biodegradable waste management is divided into diverse types such as slaughterhouse waste management, municipal solid waste management, food waste management and sewage sludge treatment. The topics elaborated in this chapter will help in gaining a better perspective about these diverse kinds of waste management.

SLAUGHTERHOUSE WASTE MANAGEMENT

Composting

Composting is an aerobic process by which organic materials are degraded through the activities of successive groups of microorganisms. Substrates vary, and can include different types of organic and inorganic wastes, sewage sludge, pig, cattle or poultry manure, garden waste and municipal solid waste. The positive effects of composts on arable soil have been reported by many authors.

An efficient and satisfactory composting process is dependent on the presence of a high microbial diversity. Different microbial communities predominate during the four consecutive phases involving mesophilic, thermotolerant, and thermophilic aerobic microorganisms. Continual change in environmental conditions in composts (temperature, pH, aeration, moisture, availability of substrates) results in stages of exponential growth and stationary phases for different organisms. Knowledge on the diversity of the microbial communities of compost is increasing rapidly through the use of molecular biology tools, such as the COMPOCHIP microarray.

Among the concerns regarding the composting process and the use of composts in agriculture and horticulture, however, is the survival and spread of animal, human and plant pathogens. Thus, any composting process must be capable of the elimination of any health risk that may be present in the end product. An improperly managed composting process can induce the proliferation and dispersion of potentially pathogenic and/or allergenic thermotolerant/thermophilic fungi and bacteria.

Two types of composting systems exist, windrow and in-vessel systems. In-vessel composting has several advantages over windrow composting, namely, that it requires less space, provides better control than windrows, and offers a high process efficiency. The higher temperatures achieved however in windrow composting have been found to result in greater bacterial and pathogen reductions. According to Noble and Roberts, the cool zones in in-vessel composting systems where there is no or little turning, are of concern for pathogen survival. Thus, windrow composting of slaughterhouse wastes is the more appropriate approach.

Composting provides an inexpensive alternative and environmentally acceptable method for the disposal of slaughterhouse wastes. The temperatures reached during the decomposition process can kill or greatly reduce most pathogens, thus reducing the chance to spread disease. Although the heat generated during composting results in a reduction of microorganism and pathogen numbers, it is not sufficient to completely sterilize the end product, thus leaving some potential for survival and (re)growth of pathogens. The levels of pathogenic bacteria remaining at the end of the composting process are dependent on the temperatures reached and the time the temperature was maintained. Legal requirements in various countries ask for a minimum temperature of 65 °C for at least 6 consecutive days or two 3-day periods >65 °C. Achieving an average temperature of 55–60 °C for 1–2 days is generally sufficient to reduce pathogenic viruses, bacteria, protozoa, and helminth ova to an acceptably low level. However, the endospores produced by spore-forming bacteria such as Clostridium and Bacillus would not be inactivated under these conditions. Typically, pH values exceeding 8 are also obtained in the composting process which also contributes to the inactivation of pathogens. Pathogens in compost can also be inactivated by the action of antibiotics produced by various microorganisms present in the compost.

Composting can potentially serve as an acceptable method of disposal of slaughterhouse wastes. However, some additional treatments would be necessary to inactivate some pathogens in the end product. Composting can also serve as a method of post- treatment of AD wastes. This post-treatment helps to ensure a more complete breakdown of organic matter, and enriches the compost in terms of the nutrients. Composting of the anaerobic digestate would in addition act to reduce pathogen levels in such products.

AH

AH represents a relatively new technology for the disposal of animal carcass material, and other infectious wastes. The process uses sodium or potassium hydroxide to catalyze the hydrolysis of biological materials (proteins, nucleic acids, carbohydrates, lipids, etc.) into a sterile aqueous solution consisting of small peptides, amino acids, sugars, and soaps. In order to accelerate the hydrolysis, the process is usually conducted with pressure and temperature. For the inactivation of microbial pathogens, carcasses must be heated to 100 °C and pressurized at 103 kpa for 3 h. To destroy prion containing material, carcasses must be heated to 150 °C and pressurized at 486 kpa for 6–8 h.

Studies investigating the inactivation of pathogens using AH are limited, however, those that exist, indicate the method to be highly successful. Neyens et al. reported that an AH process conducted at 100 °C, pH ≈10, 120 kpa and for 60 min was able to kill all pathogens. Unfortunately, it was not reported which pathogens were investigated. In a recent study, Dixon et al. investigated the use of AH at ambient temperature for the inactivation of the fish pathogens infectious salmon anaemia virus (ISAV) and Lactococcus garvieae. Both pathogens were inactivated after 48 h of treatment with 1 M NaOH at pH >13 and at room temperature. The efficacy of AH as an alternative for the treatment and disposal of infectious waste was also evaluated by testing for the destruction of specific microorganisms in a study by Kaye et al. AH was found to destroy all representative classes of potentially infectious agents.

The sterile end product of AH can be released into a sanitary sewer, in accordance with local and federal guidelines regarding pH and temperature. Studies conducted by the same authors have shown that by bubbling carbon dioxide into the hydrolysate at the end of the digestion, the pH of the end product can be reduced into the range of pH 8 or less. Considering that the hydrolysate is comprised of a mixture of single amino acids, small peptides, and fatty acids, all excellent growth nutrients for microorganisms, alternative disposal methods for this waste product should be sought. One possibility is its use as a fertilizer, however, another possibility that would require further research, but which would serve in renewable energy production, would be its use in AD. Coupled with AD, an AH unit could produce significant amounts of energy, and at the same time, safely dispose of potentially contaminated slaughterhouse waste products.

An alkaline pretreatment with NaOH was used in a study to enhance biogas production from the AD of corn stover. Different concentrations of NaOH (1–7.5% w/w) were tested and the preparations were subjected to AD. The highest biogas yields were obtained with 5% NaOH. This study shows that AH of slaughterhouse wastes followed by AD should be considered a potential route of treatment.

There are many advantages of using AH for the inactivation of disease agents in slaughterhouse wastes. These include the combination of a sterilization and digestion step into one operation, the reduction of waste volume and weight by as much as 97%, the total destruction of pathogens including prions such as BSE and the lower emission of odors or public nuisances.

AD

AD is a biological process by which organic wastes are decomposed in the absence of oxygen, producing a sludge of agricultural value, as well as biogas, which can be used to generate energy. In recent years, AD technology has been applied and used commercially in the treatment of farm, industrial, and municipal wastes. AD also represents an alternative method for the disposal of animal carcasses and slaughterhouse wastes, with the dual benefits of both eliminating waste material and producing energy, thus saving carbon credits. Concerns exist however as to whether an AD process can successfully inactivate pathogens. Often, a pasteurization step is used to help inactivate pathogens prior to AD. It is also common to include a secondary heat treatment process (e.g. composting or pasteurization) and a minimum storage period at the end of the process for the digestate as an additional measure. Swedish law for example, requires biogas plants that use animal waste to pasteurize the incoming substrate at 70 °C for 60 min prior to digestion, to ensure a hygienically acceptable end product.

The anaerobic process itself requires specific environmental conditions and requires that the bacteria and archaea involved co-operate in a close and efficient syntrophism. Digestion occurs in four major stages (hydrolysis, acidogenesis, acetogenesis, and methanogenesis) and complex polymers are degraded in a stepwise manner to yield CO_2 and CH_4.

AD can be run either under mesophilic (35 °C) or thermophilic conditions (55 °C), and the duration of the process and the effectiveness in destroying pathogens thus varies. A mesophilic process will typically run for 15–30 days, while a thermophilic process will run for 12–14 days. Mesophilic AD processes are reportedly more robust and less sensitive to changes in process parameters than are thermophilic processes, however, gas production is lower and it is more likely that the end product poses a greater pathogen risk if applied directly to the field. Methane producing microorganisms are more sensitive to temperature alterations than other bacteria, and temperature variations as low as 2 °C can reportedly have adverse effects on mesophilic AD processes.

Thermophilic AD on the other hand offers higher methane production, faster throughput, and a better inactivation of pathogens. Despite its advantages, the thermophilic digestion process needs more expensive technology, and requires a higher degree of operation and monitoring. Also, changes of only 0.5 °C can affect thermophilic AD processes.

Monteith et al. seeded a bovine enterovirus and parvovirus into liquid cattle manure, and noted that both viruses were rapidly inactivated by a thermophilic AD process, and not detected after 30 min. Under mesophilic conditions, however, the viruses were able to survive for 13 and 8 days, respectively. Spillman et al. also found that the rotavirus and coxsackievirus B5 animal viruses were rapidly inactivated by thermophilic AD. These findings indicate that viruses are inactivated during AD, but that the rate of inactivation is dependent on the virus, the temperature and the duration of digestion. Even when using a thermophilic AD process, it is advisable to use an additional heat treatment at the end of the process to fully inactivate pathogens capable of surviving AD (i.e. spore-formers). In fact, the European Commission Regulations (EC) no. 1774/2002 and no.208/2006 require a 70 °C/60 min pasteurization step of animal waste prior to landspreading of sludges. It should also be remembered however, that an additional heat treatment would still most likely not allow the inactivation of prions such as BSE.

The high lipid and protein contents of slaughterhouse waste products can cause inhibition of AD processes, as a result of an accumulation of ammonia produced from the degradation of proteins and long chain fatty acid accumulation from the breakdown of lipids. Also, lipids can form floating aggregates and foam that can cause stratification problems due to the adsorption of lipids into the biomass. However, if properly managed, AD can be conducted successfully on slaughterhouse wastes with high methane production, by allowing a progressive acclimatization of the microbial communities to an ammonia rich medium.

Pathogens

In this topic, information on the survival and inactivation of a number of specific pathogens or indicator organisms is summarized.

Escherichia Coli

Escherichia coli is a Gram negative, rod-shaped bacterium that is often found in the lower intestine of warm-blooded organisms. Most strains are harmless, although there are serotypes that cause food poisoning in humans, and strains such as O157:H7, which can cause serious illness or death in the elderly, the very young and the immuno-compromised.

A pasteurization experiment was conducted by Sahlström et al. to determine the survival rates of E. coli O157 after heat treatment. In the experiment, E. coli O157 was found to survive after 30 min at 55 °C, but not after 60 min at 55 °C. Similarly, a treatment at 70 °C for 30 min was sufficient to inactivate E. coli. Inactivation of non-pathogenic E. coli and pathogenic E. coli O157:H7 has been reported during composting of several types of waste, including animal manure and sewage sludge. The persistence however of non-pathogenic E. coli during composting and the finding of non-pathogenic E. coli in mature composts have also been reported. It is possible that in these cases the composting process employed was not conducted for a long enough time at a high enough temperature.

Laboratory scale experiments conducted by Jiang et al. have shown that E. coli 0157:H7 in bovine fecal matter are inactivated at thermophilic composting temperatures. In their study, Jiang et al. showed that large populations (104–107 CFU/g) of E. coli O157:H7 survived for 36 days during composting in a bioreactor at an external temperature of 21 °C but were undetectable after 7–14 days of composting in the bioreactor at an external temperature of 50 °C. In another study, Lung et al. investigated E. coli O157:H7 survival after cow manure composting. E. coli O157:H7 was not detected after 72 h of composting at 45 °C. Thus, a composting process conducted at a minimum temperature of 50 °C and lasting longer than a week was assumed to be able to successfully inactivate E. coli O157:H7 in manure, and could be assumed to inactivate other E. coli strains in slaughterhouse waste products.

The reductions of pathogenic and indicator bacteria (including E. coli) in animal slurry subjected to mesophilic (35 °C) and thermophilic (53 °C) AD was investigated in a study using both small-scale and full-scale reactors. At small-scale digestion at 35 °C, a T90 value (time required for a 90% reduction in number of bacteria) of 1·8 days was measured for E. coli. At 53 °C, the T90 value was reduced to 0·4 h for E. coli. Similar values were obtained at full-scale. Thus, we would hypothesize that a combination of a pasteurization step prior to AD, followed by an AD process should totally inactivate all E. coli cells present in any slaughterhouse waste material.

No information could be found concerning the use of AH to inactivate E. coli. However, considering that AH was found in a study by Kaye et al. to completely destroy all representative classes of potentially infectious agents as well as disposing of animal carcasses by solubilization and digestion, it can be assumed the process would also totally inactivate E. coli.

Salmonella

Salmonella are ubiquitous and cause illnesses such as typhoid and paratyphoid fever, and very frequently, food poisoning. All serovars of Salmonella are potentially pathogenic to both humans and animals, and the organism is spread in the environment by animal slurry, and sewage sludge. Salmonella is often used as an indicator organism to test whether a particular treatment process has been successful in the inactivation of microorganisms. Many studies have been conducted investigating the survival of Salmonella at various temperatures, for different periods of time. Generally, Salmonella is not able to survive at temperatures above 70 °C. With the exception of S. senftenberg, Salmonellae are destroyed at 56 °C for 10–20 min, although there is better heat tolerance at low water activity and in high fat foods. In a study by Paluszak et al., S. senftenberg W775 introduced into meat material was found to be eliminated by a composting process within 30 h at 68 °C. However, Salmonella has been reported to be able to re-colonize composts when temperatures are reduced near the end of the composting process if the process did not reach sufficiently high temperatures, or if the pile was not adequately aerated or turned.

Temperature is the most important factor affecting the survival of a pathogenic microorganism during AD. Studies by Plym-Forshell showed that Salmonella could be inactivated within 24 h in a thermophilic digestion process, while weeks-months were required in a mesophilic digestion process. However, in a study conducted by Gadre et al. all Salmonella species investigated were inactivated after 10 days at 37 °C. Pasteurization is frequently used prior to AD in order to reduce pathogen numbers. As it is known that Salmonella in sludge cannot withstand more than 5 min of heating at 70 °C, a pasteurization step should act as extra insurance for Salmonella inactivation from slaughterhouse wastes and in animal material after AD.

Literature does not reveal much information on the use of AH to inactivate Salmonella. A study by Mastroeni et al. found AH of crude whole Salmonella extracts reduced the lethal toxicity of the extracts on mice. Extracts were however only incubated with NaOH at a final concentration of 0.25 M for 3 h at 37 °C. This is much lower a temperature than what is used in AH units. Considering this, and that AH was found in a study by Kaye et al. to completely destroy all representative classes of potentially infectious agents as well as disposing of animal carcasses by solubilization and digestion, it can be assumed that the process would also totally inactivate Salmonella species.

Clostridium (Botulism, Black Quarter and Tetanus)

Clostridium botulinum causes the disease botulism. The disease can occur following ingestion of contaminated food, from colonization of the infant gastrointestinal tract, or from a wound infection. Vegetative cells and spores of C. botulinum can be found in soil, decaying vegetation and manure. The disease results from the production of a potent neurotoxin which is released when spores germinate.

Black quarter is an infectious and often lethal bacterial disease of sheep and cattle caused by Clostridium chauvoei. The bacteria can be found in soil and faeces, and once a pasture has become contaminated, the disease usually occurs annually in susceptible animals that graze on the pasture.

Clostridium tetani is the causal agent of tetanus, a disease characterized by muscular spasms that can lead to respiratory failure and, in up to 40% of cases, death. Spores of the bacteria can reside in soil or in the gastrointestinal tract of animals. C. tetani usually enters a host through a wound to the skin. C. botulinum, C. chauvoei and C. tetani can cause serious diseases in farm animals, resulting in suffering or even death for the animals. Spore-forming bacteria including pathogenic Clostridium species can survive pasteurization and are reportedly not affected by the AD process. Couturier & Galtier report survival of Clostridium following thermophilic digestion. The survival of pathogenic Clostridia during the biogas process results in the need for careful assessment of the risk of application of digested residues onto agriculture. A study by Bagge et al. however failed to show the survival of pathogenic clostridia after AD.

It has been reported that Clostridium can survive the composting process. In a carcass-composting environment, it has been found that anaerobic zones do develop, despite attempts to keep the entire process aerobic. These anaerobic zones allow the growth of Clostridium cells. Eventually, as decomposition occurs, and the temperatures increase, Clostridium growth will slow, and sporulation will occur.

The AH process has been shown to destroy all pathogens listed as index organisms by the State and Territorial Association on Alternative Treatment Technologies (STAATT I and STAATT II), which require a 6-log (99.9999%) reduction in vegetative agents and a 4-log (99.99%) reduction in spore-forming agents. Thus, it could be assumed that Clostridium would be inactivated through AH, and that AH would appear to be the best approach that could be taken to inactivate spore-forming clostridial pathogens.

Brucella Abortus (Brucellosis)

Bacteria of the genus Brucella are the causative agents for the zoonotic disease brucellosis. B. abortus infects cattle primarily, although it can occasionally also infect sheep. B. abortus infects the placenta and fetus of gestating cows and causes the fetus to abort. When humans however, are infected by this organism they develop a severe fever, but do not abort.

Bacteria of the genus Brucella have been reported to be very heat sensitive. Pasteurization at 63 °C for 30 min and at 72 °C for 15 sec successfully removed Brucella from contaminated raw milk, making it safe for consumption, indicating that pasteurization of animal carcasses could be used to remove the risk of disease. These findings would also suggest that composting could also inactivate Brucella. Similarly, Harada et al. reported that a temperature of 61 °C for 3 min would inactivate Brucella abortis or suis. It is thus likely that a thermophilic AD process would also inactivate this pathogen.

Again, although no literature reports regarding the survival of Brucella after a process of AH could be found based on the report of Kaye et al. it could be expected that AH would be capable of the inactivation of Brucella species.

Bacillus Anthracis

Anthrax is an acute disease caused by Bacillus anthracis. The disease affects both humans and animals, and most forms of the disease are lethal. Infections occur when spores of B. anthracis enter the body via abrasion, inhalation, ingestion, and, on rare occasions, via insect vectors. The endospores are extremely persistent in the environment and can remain viable in soils for up to 300 years.

Reuter et al. conducted an experiment to speculate on the possible fate of B. anthracis within a field scale composting system. Because of regulatory requirements, safety considerations and the field scale design of the experiment, B. anthracis spores were not used, but rather spores of B. licheniformis and B. thuringiensis were chosen. The study showed that a one millionfold inactivation of Bacillus spores was achieved by exposure to compost and elevated temperatures over time. Variation in composting microclimates within the compost pile may explain why some spores remained viable in certain parts of the compost pile. A total inactivation of Bacillus spores was not possible. However, composting may still have merit as a means of bio-containment of infected carcasses and slaughterhouse wastes, reducing and diluting the transfer of infectious spores into the environment.

Bagge et al. investigated the diversity of spore-forming bacteria in cattle manure, slaughterhouse waste and samples from biogas plants. Of particular interest in their study, were Clostridium and Bacillus. Before digested residues are reused for agriculture, biowaste must be adequately treated and proven to be hygienically safe. To reduce the risk of spreading pathogens, pasteurization of biowaste at 70 °C for an hour prior to AD is recommended. Pasteurization however will not inactivate bacterial spores. The results of Bagge et al. showed that Bacillus species seemed to pass the biogas process from biowaste to digested residues relatively unaffected. Due to their survival of pasteurization and the AD process, it is important to assess the risk of spreading digested residues from biogas plants containing Bacillus spores onto agricultural fields, and to take the necessary precautions.

The efficacy of AH for the inactivation of pure cultures of bacteria, including B. subtilis, for the disposal of infectious animal waste was evaluated in a study by Kaye et al. The tissue digester was operated at 110–120 °C and approximately 103 kpa for 18 h. Samples were removed after a cooling phase at 50 °C. The process was found to completely destroy B. subtilis. It can thus be expected that AH would be capable of the inactivation of the more deadly pathogen, B. anthracis.

Mycobacterium Bovis (Beef Tuberculosis)

Mycobacterium bovis is the causative agent of bovine tuberculosis, a chronic disease of cattle and other species. The disease can cause major losses in animal production and is a potentially debilitating zoonosis. M. bovis is capable of jumping the species barrier, and causing tuberculosis in humans, a disease normally caused by M. tuberculosis. Normally, M. bovis is transmitted to humans via infected milk, although it can also spread via aerosol droplets. However, actual infections in humans are rare, because of control measures put in place. Studies by Croshaw et al. showed that M. bovis was destroyed by the pasteurization of milk at 71.7 °C for 15 sec, and at 62.8 °C for 30 min.

Far more work has been conducted investigating the survival of Mycobacterium avium subsp. paratuberculosis after composting than M. bovis. M. avium subsp. paratuberculosis is a member of the same genus, and the causal agent of paratuberculosis, an intestinal infection of domestic and wild ruminants. Grewel et al. showed that after 3 days of composting at 55 °C, M. avium subsp. paratuberculosis were no longer detectable. Also, Gobec et al. showed that M. avium subsp. paratuberculosis was not able to be isolated 24 h after the start of composting. However, a report by Hahesy indicated that farm yard (composted) manure must be exposed to a mean temperature of 60–70 °C during the composting process for 3 weeks to destroy M. bovis bacilli. The authors concluded that as the majority of solid dung heaps do not reach this high a temperature, composted manure cannot necessarily be considered safe. These findings appear to be contradictory of the results of related Mycobacterium studies, and of the findings of pasteurization studies by Croshaw et al.

No studies investigating M. bovis in AD could be found upon intensive literature searches. In a mesophilic AD study by Olsen et al. M. paratuberculosis was detected and cultured up until 28 days of digestion. Under thermophilic conditions, however, M. paratuberculosis was inactivated after 24 h. In a study by Slana et al. viable M. avium subsp. paratuberculosis cells were detected using culture from fermenters up to 2 months. Therefore, the safety of the final products of digestion used for fertilization concerning the presence of Mycobacterium species cannot be certain.

Kaye et al. tested the survival of organisms after AH. All organisms tested, including M. bovis were completely destroyed by the AH process.

Erysipelothrix Rhusiopathiae

Erysipelothrix rhusiopathiae is found all over the world and has been isolated from a variety of vertebrates as well as from soil. E. rhusiopathiae is primarily considered an animal pathogen, causing a disease known as erysipelas in turkeys, pigs, birds, sheep, fish, and reptiles.

Erysipelothrix cultures are reportedly destroyed by exposure to moist heat at 55 °C for 10 min. A study conducted by Morrow et al. investigated composting as an approach to treat swine mortalities. The authors placed cultures of E. rhusiopathiae in compost piles. The temperatures reached in the composting process were high enough to destroy E. rhusiopathiae, and the authors concluded that properly produced composts did not pose a risk for transmitting disease.

Olsen and Larsen published the results of experiments investigating the reductions of pathogenic and indicator bacteria in animal slurry subjected to mesophilic (35 °C) and thermophilic (53 °C) AD. Under mesophilic AD conditions, a T90 value of 1·8 days was obtained for E. rhusiopathiae, while at 53 °C, the T90 value was reduced to 1–2 h. A study conducted by Han et al. investigated the effects of an anaerobic lagoon fermentation and an autothermal thermophilic aerobic digestion on bacterial pathogens contained in raw swine manure. E. rhusiopathiae was present in the raw manure, however was no longer detectable following a mesophilic anaerobic lagoon fermentation. Manure treated by thermophilic aerobic digestion also did not yield detectable levels of E. rhusiopathiae. These findings suggest that AD processes would be successful in the inactivation of E. rhusiopathiae.

Although no literature reports regarding the survival of E. rhusiopathiae after a process of AH could be found, the report of Kaye et al. would indicate that AH should be capable of the inactivation of E. rhusiopathiae.

BSE

BSE, also known as mad cow disease, is a relatively new disease that primarily affects cattle. BSE can also cause a corresponding disease in humans-Creutzfeldt-Jakob disease. There is still much controversy regarding the causes of BSE and Creutzfeldt-Jakob disease, however the most common belief is that the infectious agents are prions, an abnormal form of a type of protein. However, the prion hypothesis has also been challenged and an autoimmune response theory has been postulated.

Currently, the public considers BSE to be the greatest concern to any bovine-based product. The risk, however, of spreading BSE via composting of catering wastes in the UK has been shown to be 'remote', because there are many controls in place for keeping the disease from entering the food system, and hence the food residuals stream in the UK today. In the slaughterhouse, TSE management aims to prevent infected material from entering the food and feed chains. According to EU legislation, animals suspected of TSE infection are separated and safely disposed of. Prions such as BSE are more resistant to heat than many viruses. In fact, BSE infected material remains infected after cooking, rendering and long periods of incubation in the soil. According to Rohwer, less than a 0.5 log (70%) destruction of scrapie agent was seen after 60 min at 60 °C and 80 °C. It can therefore be assumed that a standard composting process whereby the temperature was maintained at 60 °C for 2 days would not reduce BSE infectivity.

TSEs have also been reported to survive the operational temperatures at which AD are conducted. Studies by Topper et al. showed AD under both mesophilic and thermophilic conditions to be incapable of reducing or eliminating BSE. In a study conducted by Brown et al. infected brain material was heated to 600 °C. Despite brains being totally ashed, when reconstituted with saline to their original weights, ashed brain material was able to transmit disease to 5 of 35 inoculated hamsters.

Apart from incineration, AH is the only effective method known for the destruction of prion material. A study carried out by the Institute of Animal Health at the University of Edinburgh, investigated the ability of AH to destroy BSE prions grown in the brains of mice. Infected mice heads were subjected to AH for either 3 or 6 h, and after neutralization of the hydrolysates, aliquots were injected into mice. Evidence of TSE was found in some mouse brains of mice injected with hydrolysate taken from the 3 h digestion, but significantly, no disease was found in the brains of mice injected with hydrolysate from the 6 h digestion. Studies conducted by Murphy et al. also showed scrapie to be inactivated by AH.

Aphto Virus

Foot-and-mouth disease (FMD) is an infectious and sometimes fatal viral disease caused by the Aphto virus that affects domesticated and wild ruminants and pigs. Most infected animals eventually recover but often develop sequelae. The highly contagious nature of the virus and the severity of economic impacts associated with disease have resulted in FMD being labeled as the most important disease limiting the trade of animals and animal products throughout the world.

The virus can survive for weeks or months in refrigerated internal organs, bone marrow, lymph and hemal nodes, glands, and residual blood, however can be inactivated within 3 days in skeletal muscle after slaughter due to reduced pH. The FMD virus can be inactivated using heat, however, the amount of virus inactivation is determined by the temperature, and by access of the heat to the virus infected material. Blackwell et al. reported that heating to an internal temperature of 58.3 °C inactivated FMD in minced beef containing virus-infected lymph nodes, while temperatures of up to 79.4 °C were required for minced meat in nylon tubes. The American Association of Food Hygiene Veterinarians reported that core temperatures of 93 °C were needed to inactivate the FMD virus in heart muscle.

Guan et al. investigated the inactivation and degradation of FMD virus during composting of infected pig carcasses. FMD virus infected pig carcasses were composted in a mixture of chicken manure and wood shavings, and compost temperatures reached 50 °C and 70 °C by days 10 and 19, respectively. Results indicated that the virus was inactivated in specimens in compost by day 10 and the viral RNA was degraded in skin and internal organ tissues by day 21.

Inactivation experiments conducted by Turner et al. in contrast, showed the FMD virus to be remarkably thermally stable, surviving with little or no loss of titre at 55 °C and 60 °C. Even at 65 °C, the virus did not drop to below detectable levels within 5 min in a slurry. From these studies, the authors recommended treatment of slurry and water-containing FMD virus in full-scale mobile treatment plant operating at a minimum of 70 °C.

As with other viruses, the bone marrow of infected animal carcasses can contain high loadings of FMD virus. Therefore, to effectively inactivate FMD virus from animal waste material, a composting process where the temperature was maintained at greater than 65 °C for at least 2 days would need to be used. According to Gale et al., a composting temperature of 60 °C for 2 days would give at least a 6-log reduction of all exotic pig viruses, including FMD, even in the bone marrow of whole pig carcasses discarded for composting.

No information could be found regarding the survival of FMD virus after either mesophilic or thermophilic AD. Considering the heat stability of the virus, it could be predicted that AD would not successfully inactivate FMD virus. However, the combination of a pasteurization step and an AD process could be expected to be successful in the inactivation of the virus.

No literature reports regarding the survival of FMD virus after a process of AH could be found. As the report of Kaye et al. states that the process was found to completely destroy all representative classes of potentially infectious agents, it can be expected that AH would be capable of the inactivation of this virus.

Lyssa Virus (Rabies)

Rabies is a disease caused by the lyssa virus, and is an acute encephalitis in warm-blooded animals. In humans, rabies is almost always fatal if post-exposure prophylaxis is not administered prior to the onset of severe symptoms. The economic impact of the disease is significant, with the virus causing death of livestock in some countries.

The rabies virus is able to be inactivated by heat and thus a pasteurization step at 70 °C should be considered successful in rabies inactivation. No information was able to be found on composting inactivation of the virus, but based on the findings of Kopcha et al, it would be likely that composting would also successfully inactivate the virus. Studies by Kissling and Reese investigated virus stability at different temperatures. Rabies tissue culture-adapted virus in serum-free medium was reported to be sensitive to heat inactivation. At 56 °C, almost all virus was inactivated within 30 min, while at 37 °C, more than 99 % of the virus was inactivated within 24 h. At 37 °C, no virus was viable after 32–48 h.

No information could be found upon extensive literature searching on the survival of the rabies virus after an AD process. Considering the heat sensitivity of the virus, it is likely that it would be inactivated after a thermophilic AD, but it is not certain whether a mesophilic AD process would have the same effect. When combined with a pasteurization step however, one could hypothesize that the virus would be inactivated.

Again, no literature reports regarding the survival of the rabies virus after a process of AH could be found. Based on the findings of Kaye et al. it would however be expected that inactivation of the virus would result from AH.

African Swine Fever Virus

African swine fever (ASF) is a viral disease (genus Asfivirus) which causes a devastating haemorrhagic fever of pigs with mortality rates approaching 100%. Virus is excreted in the faeces of infected pigs, and as faeces mostly are transferred to a slurry store prior to the infection being identified, viral contamination of large quantities of pig slurry can occur inadvertently.

ASF virus has been reported to be quite resistant to inactivation, and can survive many freeze-thaw cycles. It is also resistant to changes in pH, and certain strains are resistant to complete inactivation at pH values between 4 and 13. The virus is however very sensitive to drying, and can be inactivated by lipid solvents because of its lipid envelope.

ASF virus is however heat sensitive. In studies conducted by Plowright and Parker, ASF virus was shown to be inactivated (5 log inactivation) by 90 min at 56 °C. However, work done by Turner et al. indicated the virus to be inactivated at 50 °C in 30 min, at 56 °C within 90 sec and within seconds at 60 °C.

The findings of Turner et al. described above, as well as projected costs of treatment for virus inactivation resulted in the authors deciding that heat was the most effective means of ASF virus inactivation. This is because heat treatment was found to be very rapid and consistent at a temperature of 56 °C, while alkali inactivation was slower. Also, heat can be recovered, thus reducing costs. These authors designed a pilot scale treatment plant to treat pig slurry at a minimum temperature of 65 °C, and maintained this temperature in at least 99.99% of the material for a minimum of 5 min. This was done so to provide a reasonable safety margin for virus inactivation.

The bone marrow of infected pig carcasses contains high loadings of ASF virus. When considering the composting of an animal carcass, the times and temperature of the process must be carefully considered. It has been estimated that a sphere of diameter 40 cm (which would be similar to a large leg or carcass of pork containing the bone) would require 40 h to reach a temperature of 56 °C at the centre when the surrounding composting temperature is 60 °C. Therefore, a composting process with a temperature of 60 °C maintained for 2 days would give at least a 6-log reduction of all exotic pig viruses, including ASF virus, even in the bone marrow of whole pig carcasses. As studies have shown ASF virus to be inactivated within seconds at 60 °C, it would appear that a correctly performed composting process should be successful in inactivating ASF virus.

No information could be found regarding AD and the survival rates of ASF virus subjected to the process. While no predictions can be made as to the survival of the virus in a mesophilic process, it could be predicted that a thermophilic process may successfully inactivate the virus.

A 1% solution of both NaOH or $Ca(OH)_2$ was found to be able to result in a 4 fold inactivation of the virus at 4 °C within 150 sec. More dilute solutions (0.5%) of both chemicals were however not found to be able to inactivate virus at 22 °C in 5 min. It could be expected that a AH process, which uses alkaline solution as well as heat to inactivate microbial organisms would be successful in the inactivation of ASF virus.

Phlebo Virus (Rift Valley Fever)

Rift Valley fever (RVF) is a viral zoonosis resulting in fever. The disease is caused by the RVF virus, a member of the Phlebovirus genus. It is spread by the bite of infected mosquitoes, or by exposure to tissues or blood of infected animals. Humans infected with the virus have self-limited febrile illness, but retinal degeneration (5–10%), hemorrhagic fever (<1%), or encephalitis (<1%) can also develop. The RVF virus is susceptible to low pH (≤ 6.2), lipid solvents and detergents, as well as solutions of sodium or calcium hypochlorite with residual chlorine content greater than 5000 ppm. In a neutral or alkaline pH, mixed with serum or other proteins, the virus can survive for as long as four months at 4 °C and eight years below 0 °C. The virus is also quickly destroyed by pH changes in decomposing carcasses. Thus, although no literature can be found on AH of animals infected with RVF virus, it would appear that AH would not be a recommended method of treatment of animal carcasses in order to inactivate the virus.

Unfortunately, no information could be found on composting or AD with respect to the inactivation of RVF virus, although Flory et al. report that composting approaches can be used to treat RVF diseased animals, thus preventing disease outbreaks.

Cysticercus Bovis (Beef Measles)

Bovine cysticercosis is a zoonotic disease found worldwide affecting the muscles of cattle. The disease is most prevalent in countries where poor sanitation practices on cattle farms are common and where cultural habits include eating undercooked meat. It does not represent a serious human health risk. Bovine cysticercosis is caused by Cysticercus bovis, the larval stage of Taenia saginata, a species of human tapeworm. Humans can acquire intestinal tapeworm infections by the consumption of beef containing the parasite. Disease in cattle is initiated by the ingestion of materials contaminated with tapeworm eggs originating from human feces larvae of a human tapeworm. Cattle become infected with bovine cysticercosis by ingesting materials contaminated with tapeworm eggs originating from human faeces.

A study was conducted by Hughes et al. to investigate the survival of T. saginata ova after AD at 35 °C for 10 days followed by lagooning for 15 days. Results indicated that viability and infectivity were completely destroyed. The authors also showed that a 3 h treatment at 55 °C in crude sludge resulted in a 99 % reduction of ova infectivity. It could thus be assumed that a 70 °C pasteurization step prior to AD treatment, as is the regulation in Europe, would successfully inactivate bovine cysticercosis.

Another study by Morris et al. showed 100% efficacy of sanitation of T. saginata at 35 °C. A thorough review of the literature has failed to find studies whereby the inactivation of T. saginata by composting was investigated. However, the US Environmental Protection Agency (Part 503) sets standards for the destruction and concentration of pathogens in composted biosolids. These standards are commonly used as marketplace specifications for compost sold for public use. By maintaining these standards, various harmful pathogens should be destroyed. According to the standards, T. saginata does not survive for more than a few minutes at 55 °C. Thus, a standard composting procedure would successfully inactivate this organism.

No information could be obtained regarding the inactivation efficiency of AH on T. saginata. However, Kaye et al. showed Giardia cysts to be completely destroyed by AH, an indication that slaughterhouse waste material infected by bovine cysticercosis might also successfully be treated by AH.

MUNICIPAL SOLID WASTE MANAGEMENT

Solid Waste Management may be defined as the discipline associated with the control of generation, collection, storage, transfer and transport, processing and disposal of solid wastes in a manner that is in accord with the best principles of public health, economics, engineering, conservation, aesthetics and other environmental considerations.

The most commonly recognized methods for the final disposal of solid wastes are:

- Dumping on land,
- Dumping in water,
- Ploughing into the soil,
- Incineration.

Municipal Solid Wastes

Municipal solid waste includes commercial and domestic wastes generated in municipal or notified areas in either solid or semi-solid form excluding industrial hazardous wastes but including treated bio-medical wastes.

Collection of Municipal Solid Wastes

Littering of municipal solid waste shall be prohibited in cities, towns and in urban areas notified by the State Governments. To prohibit littering and facilitate compliance, the following steps shall be taken by the municipal authority, namely:

- Organizing house-to-house collection of municipal solid wastes through any of the methods, like community bin collection (central bin), house-to-house collection, collection on regular pre-informed timings and scheduling by using musical bell of the vehicle.

- Devising collection of waste from slums and squatter areas or localities including hotels, restaurants, office complexes and commercial areas.
- Wastes from slaughter houses, meat and fish markets, fruits and vegetable markets, which are biodegradable in nature, shall be managed to make use of such wastes.
- Bio-medical wastes and industrial wastes shall not be mixed with municipal solid wastes and such wastes shall follow the rules separately specified for the purpose.
- Collected waste from residential and other areas shall be transferred to community bin by hand-driven carts or other small vehicles.
- Construction or demolition wastes or debris shall be separately collected and disposed off following proper norms. Similarly, wastes generated at dairies shall be regulated in accordance with the State laws.
- Waste (garbage, dry leaves) shall not be burnt.
- Stray animals shall not be allowed to move around waste storage facilities or at any other place in the city or town.

Storage of Municipal Solid Wastes

Municipal authorities shall establish and maintain storage facilities in such a manner, as they do not create unhygienic and unsanitary conditions around it. Following criteria shall be taken into account while establishing and maintaining storage facilities, namely:

- Storage facilities shall be created and established by taking into account quantities of waste generation in a given area and the population densities. A storage facility shall be so placed that it is accessible to users.
- Storage facilities to be set up by municipal authorities or any other agency shall be so designed that wastes stored are not exposed to open atmosphere and shall be aesthetically acceptable and user-friendly.
- Storage facilities or 'bins' shall have 'easy to operate' design for handling, transfer and transportation of waste. Bins for storage of bio-degradable wastes shall be painted green, those for storage of recyclable wastes shall be painted white and those for storage of other wastes shall be painted black.
- Manual handling of waste shall be prohibited. If unavoidable due to constraints, manual handling shall be carried out under proper precaution with due care for safety of workers.

Processing of Municipal Solid Wastes

Municipal authorities shall adopt suitable technology or combination of such technologies to make use of wastes so as to minimize burden on landfill. Following criteria shall be adopted, namely:

- The biodegradable wastes shall be processed by composting, vermicomposting, anaerobic digestion or any other appropriate biological processing for stabilization of wastes.

- Mixed waste containing recoverable resources shall follow the route of recycling.
- Incineration with or without energy recovery can also be used for processing wastes in specific cases.

Municipal authority or the operator of a facility wishing to use other state-of-the-art technologies shall approach the Central Pollution Control Board to get the standards laid down before applying for grant of authorization.

Disposal of Municipal Solid Wastes

Land filling shall be restricted to non-biodegradable, inert waste and other waste that are not suitable either for recycling or for biological processing. Land filling shall also be carried out for residues of waste processing facilities as well as pre-processing rejects from waste processing facilities. Land filling of mixed waste shall be avoided unless the same is found unsuitable for waste processing. Under unavoidable circumstances or till installation of alternate facilities, land filling shall be done following proper norms.

Managing Non-Biodegradable Solid Waste (NBDSW)

Non-biodegradable solid waste (NBDSW) or refuse is a carpet word. It covers a variety of materials ranging from asbestos to Zinc batteries. Polythene and its related compounds are the most commonly found solid waste materials in urban environs. Many non-biodegradable solid waste materials are known to cause considerable environmental hazards when released into land, water and atmosphere.

Coastal Environment and Social Waste Management

Solid waste related problems prevail more in megalopolis and the dangers reach great heights in coastal cities. Solid wastes of domestic and industrial units are considered major pollutants of coastal regions of India.

Waste Management at Source

Source of waste generation	Action to be taken
Household	• Not to throw any solid waste in the neighbourhood, on the streets, open spaces, and vacant lands, into the drains or water bodies. • Keep food waste/biodegradable waste in a non corrosive container with a cover (lid). Keep dry, recyclable waste in a bin or bag or a sack. • Keep domestic hazardous waste if and when generated separately for disposal at specially notified locations.
Multi-storeyed buildings commercial complexes private societies	• Provide separate community bin or bins large enough to hold food/biodegradable waste and recyclable waste generated in the building or society. • Direct the members of the association to deposit their waste in community bin.
Slums	• Use community bins provided by local body for deposition of food and biodegradable waste.

Shops, offices, institutions, etc.	• If situated in a commercial complex, deposit the waste in bins provided by the association.
Hotels & restaurants	• The container used should be strong, not more than 100 litre in size, should have a handle on the top or handles on the sides and a rim at the bottom for easy handling.
Vegetable & Fruit Markets	• Provide large containers, which match with transportation system of the local body. • Shop keepers not to dispose of the waste in front of their shops or open spaces. Deposit the waste as and when generated into the large container placed in the market.
Meat & fish markets	• Not to throw any waste in front of their shops or open spaces around. Keep non-corrosive container/containers not exceeding 100-litre capacity with lid handle and the rim at the bottom and deposit the waste in the said containers as and when generated. • Transfer the contents of this container into a large container provided by the association.
Street food vendors	• Not to throw any waste on the street, pavement or open spaces. Keep bin or bag for the storage of waste that generates during street vending activity. • Preferably have arrangements to affix the bin or bag with the hand–cart used for vending.
Marriage halls, community halls, kalyanamandapas	• Not to throw any solid waste in their neighbourhood, on the streets, open spaces, and vacant lands, into the drains or water bodies. • Provide a large container with lid which may match with the transportation system of the local body and deposit all the waste generated in the premises in such containers.
Hospitals, Nursing homes, etc.	• Not to throw any solid waste in their neighbourhood, on the streets,open spaces, and vacant lands, into the drains or water bodies. • Not to dispose off the biomedical waste in the municipal dust bins or other waste collection or storage site meant for municipal solid waste. • Store the waste as per the directions contained in the government of India, Ministry of Environment Biomedical Waste (Management & Handling) Rules, 1998.
Construction/demolition waste	• Not to deposit construction waste or debris on the streets, footpaths, pavements, open spaces, water bodies etc. • Store the waste within the premises or with permission of the authorities just outside the premises without obstructing the traffic preferably in a container if available through the local body or private contractors.
Garden waste	• Compost the waste within the garden, if possible Trim the garden waste once in a week on the days notified by the local body. • Store the waste into large bags or bins for handing over to the municipal authorities appointed for the purpose on the day of collection notified.

Items that can be Recycled or Reused

Paper

- Old copies

- Old books
- Paper bags
- Newspapers
- Old greeting cards
- Cardboard box

Plastic

- Containers
- Bags
- Sheets

Glass and Ceramics

- Bottles
- Plates
- Cups
- Bowls

Miscellaneous

- Old cans
- Utensils
- Clothes
- Furniture

Role of NGOs

During the recent years, NGOs (non-governmental organizations) have taken up initiatives to work with local residents to improve sanitation. They have been playing an active role in organizing surveys and studies in specified disciplines of social and technological sciences. In the field of garbage management, such studies are useful in identifying areas of commercial potentials to attract private entrepreneurs. They can play an important role in segregation of waste, its collection and handling over to local authorities.

They are all successfully creating awareness among the citizens about their rights and responsibilities towards solid waste and the cleanliness of their city. These organizations promote environmental education and awareness in schools and involve communities in the management of solid waste.

The NGO Programmes

- Create mass awareness, ensuring public participation in segregation of recyclable material and storage of waste at source.
- Provide employment through organizing door-to-door collection of waste.
- Ensure public participation in community based primary collection system.
- Encourage minimization of waste through in-house backyard composting, vermicomposting and biogas generation.

Urban poverty is inextricably linked with waste. In India alone, over a million people find livelihood opportunities in the area of waste; they are engaged in waste collection (popularly known as rag picking) and recycling through well-organized systems. Substantial populations of urban poor in other developing countries also earn their livelihood through waste. It is important to understand issues of waste in this context. The informal sector dealing with waste is engaged in various types of work like waste picking, sorting, and recycling at the organized level, door-to-door collection, composting and recycling recovery.

Occupational Hazards Associated with Waste Handling Infections

- Skin and blood infections resulting from direct contact with waste, and from infected wounds.
- Eye and respiratory infections resulting from exposure to infected dust.
- Different diseases that results from the bites of animals feeding on the waste.
- Intestinal infections that are transmitted by flies feeding on the waste.

FOOD WASTE MANAGEMENT

Food waste is one of the most prominent waste streams across Middle East, especially in GCC region. The mushrooming of hotels, restaurants, fast-food joints and cafeterias in the Middle East region has resulted in the generation of huge quantities of food wastes. The proportion of food waste in municipal waste stream is gradually increasing and hence a proper food waste management strategy needs to be devised to ensure its eco-friendly and sustainable disposal in the Middle East.

Food waste is an untapped energy source that mostly ends up rotting in landfills, thereby releasing greenhouse gases into the atmosphere. Food waste includes organic wastes generated in hotels, restaurants, canteens, cafeterias, shopping malls and industrial parks in the form of leftover food, vegetable refuse, stale cooked and uncooked food, meat, teabags, napkins, extracted tea powder, milk products etc. It is difficult to treat or recycle food waste since it contains high levels of sodium salt and moisture, and is mixed with other waste during collection.

Food waste can be recycled by two main pathways:

- Composting: A treatment that breaks down biodegradable waste by naturally occurring micro-organisms with oxygen, in an enclosed vessel or tunnel or pit.
- Anaerobic digestion or biogas technology: A treatment that breaks down biodegradable waste in the absence of oxygen, producing a renewable energy (biogas) that can be used to generate electricity and heat.

Composting

Composting provides an alternative to landfill disposal of food waste, however it requires large areas of land, produces volatile organic compounds and consumes energy. Compost is organic material that can be used as a soil amendment or as a medium to grow plants. Mature compost is a stable material with a content called humus that is dark brown or black and has a soil-like, earthy smell. It is created by: combining organic wastes (e.g., yard trimmings, food wastes, manures) in proper ratios into piles, rows, or vessels; adding bulking agents (e.g., wood chips) as necessary to accelerate the breakdown of organic materials; and allowing the finished material to fully stabilize and mature through a curing process.

Anaerobic Digestion

Anaerobic digestion has been successfully used in several European and Asian countries to stabilize food wastes, and to provide beneficial end-products. Sweden, Austria, Denmark, Germany and England have led the way in developing new advanced biogas technologies and setting up new projects for conversion of food waste into energy. The relevance of biogas technology lies in the fact that it makes the best possible utilization of various organic wastes as a renewable source of clean energy. A biogas plant is a decentralized energy system, which can lead to self-sufficiency in heat and power needs, and at the same time reduces environmental pollution.

Of the different types of organic wastes available, food waste holds the highest potential in terms of economic exploitation as it contains high amount of carbon and can be efficiently converted into biogas and organic fertilizer. Food waste can either be utilized as a single substrate in a biogas plant, or can be co-digested with organic wastes like cow manure, poultry litter, sewage, crop residues, abattoir wastes etc.

SEWAGE TREATMENT

Sewage or municipal waste should not be directly passed into rivers, streams and other water bodies because it not only contains human excreta and other organic wastes but a number of pathogenic microbes. It is made less polluting by passing it through sewage treatment plants (STPs).

Here heterotrophic microbes naturally present in sewage carry out the process of decomposition. There are three stages of this treatment; primary, secondary and tertiary. Primary treatment is physical, secondary biological and tertiary chemical. Waste water can be passed into rivers after secondary treatment.

Primary or Physical Treatment

It is the process of removal of small and large, floating and suspended solids from sewage through two processes of filtration and sedimentation. First floating and suspended matter is removed through sequential filtration with progressively smaller pore filters. The filtrate is then kept in large open settling tanks where grit (sand, silt, small pebbles) settles down.

Aluminium or iron sulphate is added in certain places for flocculation and settling down of solids. The sediment is called primary sludge while the supernatant is called effluent. The primary sludge traps a lot of microbes and debris. It is subjected to composting, land fill or anaerobic digestion to produce biogas and manure.

Secondary or Biological Treatment

There are several methods of secondary treatment, e.g., oxidation tanks, trickling filter system and activated sludge system. In activated sludge system, the primary effluent is taken to aeration tanks. Liquid is constantly agitated mechanically. Air is also pumped into it. Some activated sludge of the previous operation is inoculated to hasten decomposition. A large number of aerobic heterotrophic microbes grow in the aeration tank.

They include bacteria of different types, some filamentous fungi, yeasts and protozoan. They form floes. Floes are masses of bacteria held together by slime and fungal filaments to form mesh like structures. The microbes digest a lot of organic matter, converting it into microbial biomass and releasing a lot of minerals.

This reduces Biochemical Demand or BOD. As the BOD of the waste matter is reduced to 10-15% of raw sewage, it is passed into settling tank. In settling tank, the bacterial floes are allowed to undergo sedimentation. The effluent or supernatant is generally passed into natural waters like rivers and streams. It can also be treated chemically to further purify it.

The sediment of settling tank is called activated sludge. A part of it is used as inoculum in aeration tanks. The remaining is passed into a large tank called anaerobic sludge digesters. They are designed for continuous operation. The aerobic microbes present in the sludge get killed. Anaerobic microbes digest the organic mass as well as aerobic microbes (bacteria and fungi) of the sludge.

They are of two types, non-methanogenic and methanogenic. Methanogenic bacteria produce marsh gas which is a mixture of gases containing methane, H_2S and CO_2. The mixture also called biogas is inflammable and is a source of energy. The spent sludge can be used as manure or part of compost.

Biological Sewage Treatment Plant

Raw sewage is passed to an aeration chamber and air is supplied through diffusers. This promotes the action of aerobic bacteria which break down the sewage into carbon di-oxide, water and inorganic waste. Figure below shows a typical biological sewage treatment plant.

Sewage then passes into the clarification or settling chamber. Any solids that settle out are returned via an air lift to the aeration chamber which ensures that they are fully broken down. The sample applies to any surface scum. A small vane type air compressor supplies the air for air diffusers and air lift.

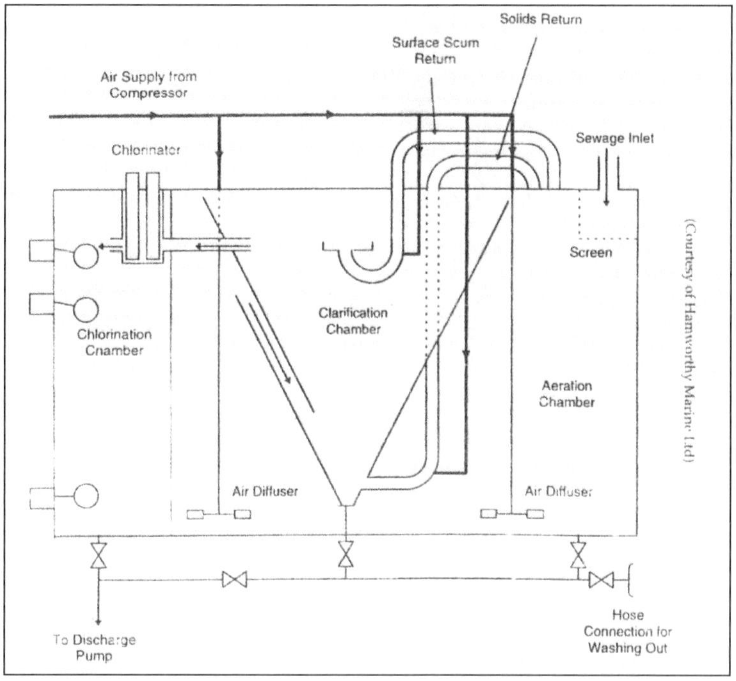

The clear liquid then passes through the chlorinator where the liquid is disinfected, into the chlorination chamber. The chamber has float switches, which control the discharge pump, and a high level alarm. Although the sewage treatment plant runs automatically, without regular maintenance the unit will not function properly and anaerobic bacteria may promote the formation of hydrogen sulphide and methane, both of which are hazardous.

The chambers need to be cleaned out occasionally to remove any accumulated matter. The aeration diffusers should be checked to ensure they are clear and that air is bubbling from them. The air lift returns should also be checked to make sure they function correctly. These usually have a clear plastic pipe so that the sludge can be seen returning to the aeration chamber. The internal tank coating should be inspected for any signs of cracking or blistering.

When cleaning out a sewage unit rubber gloves and a mask should be worn. After overhaul the external surfaces of the unit and surroundings are to be washed down with disinfectant. Hands should also be thoroughly scrubbed and overalls washed.

SEWAGE SLUDGE TREATMENT

Sewage sludge treatment describes the processes used to manage and dispose of sewage sludge produced during sewage treatment. Sludge is mostly water with lesser amounts of solid material removed from liquid sewage. Primary sludge includes settleable solids removed during primary treatment in primary clarifiers. Secondary sludge separated in secondary clarifiers includes treated sewage sludge from secondary treatment bioreactors.

Sludge treatment is focused on reducing sludge weight and volume to reduce disposal costs, and on reducing potential health risks of disposal options. Water removal is the primary means of weight and volume reduction, while pathogen destruction is frequently accomplished through heating during thermophilic digestion, composting, or incineration. The choice of a sludge treatment method depends on the volume of sludge generated, and comparison of treatment costs required for available disposal options. Air-drying and composting may be attractive to rural communities, while limited land availability may make aerobic digestion and mechanical dewatering preferable for cities, and economies of scale may encourage energy recovery alternatives in metropolitan areas.

Energy may be recovered from sludge through methane gas production during anaerobic digestion or through incineration of dried sludge, but energy yield is often insufficient to evaporate sludge water content or to power blowers, pumps, or centrifuges required for dewatering. Coarse primary solids and secondary sewage sludge may include toxic chemicals removed from liquid sewage by sorption onto solid particles in clarifier sludge. Reducing sludge volume may increase the concentration of some of these toxic chemicals in the sludge.

Biosolids

"Biosolids" is a term often used in wastewater engineering publications and public relations efforts by local water authorities when they want to put the focus on reuse of sewage sludge, after the sludge has undergone suitable treatment processes. In fact, biosolids are defined as organic wastewater solids that can be reused after stabilization processes such as anaerobic digestion and composting. The term "biosolids" was introduced by the Water Environment Federation in the U.S. in 1998. However, some people argue that the term is a form of "propaganda" (or at least a euphemism) with the aim to hide the fact that sewage sludge may also contain substances that could be harmful to the environment when the treated sludge is applied to land, for example environmental persistent pharmaceutical pollutants.

Treatment Processes

Thickening

A sewage sludge thickener.

Thickening is often the first step in a sludge treatment process. Sludge from primary or secondary clarifiers may be stirred (often after addition of clarifying agents) to form larger, more rapidly settling aggregates. Primary sludge may be thickened to about 8 or 10 percent solids, while secondary sludge may be thickened to about 4 percent solids. Thickeners often resemble a clarifier with the addition of a stirring mechanism. Thickened sludge with less than ten percent solids may receive additional sludge treatment while liquid thickener overflow is returned to the sewage treatment process.

Dewatering

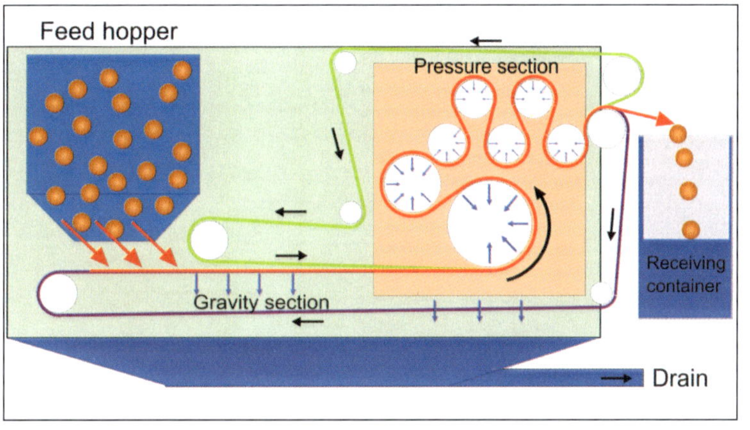

Schematic of a belt filter press to dewater sewage sludge. Filtrate is extracted initially by gravity, then by squeezing the cloth through rollers.

Water content of sludge may be reduced by centrifugation, filtration, and/or evaporation to reduce transportation costs of disposal, or to improve suitability for composting. Centrifugation may be a preliminary step to reduce sludge volume for subsequent filtration or evaporation. Filtration may occur through underdrains in a sand drying bed or as a separate mechanical process in a belt filter press. Filtrate and centrate are typically returned to the sewage treatment process. After dewatering sludge may be handled as a solid containing 50 to 75 percent water. Dewatered sludges with higher moisture content are usually handled as liquids.

Sidestream Treatment Technologies

Sludge treatment technologies that are used for thickening or dewatering of sludge have two products: the thickened or dewatered sludge, and a liquid fraction which is called sludge treatment liquids, sludge dewatering streams, liquors, centrate (if it stems from a centrifuge), filtrate (if it stems from a belt filter press) or similar. This liquid requires further treatment as it is high in nitrogen and phosphorus, particularly if the sludge has been anaerobically digested. The treatment can take place in the sewage treatment plant itself (by recycling the liquid to the start of the treatment process) or as a separate process.

Phosphorus Recovery

One method for treating sludge dewatering streams is by using a process that is also used for phosphorus recovery. Another benefit for sewage treatment plant operators of treating sludge dewatering streams for phosphorus recovery is that it reduces the formation of obstructive struvite scale in pipes, pumps and valves. Such obstructions can be a maintenance headache particularly for biological nutrient removal plants where the phosphorus content in the sewage sludge is elevated. For example, the Canadian company Ostara Nutrient Recovery Technologies is marketing a process based on controlled chemical precipitation of phosphorus in a fluidized bed reactor that recovers struvite in the form of crystalline pellets from sludge dewatering streams. The resulting crystalline product is sold to the agriculture, turf and ornamental plants sectors as fertiliser under the registered trade name "Crystal Green".

Digestion

Many sludges are treated using a variety of digestion techniques, the purpose of which is to reduce the amount of organic matter and the number of disease-causing microorganisms present in the solids. The most common treatment options include anaerobic digestion, aerobic digestion, and composting. Sludge digestion offers significant cost advantages by reducing sludge quantity by nearly 50% and providing biogas as a valuable energy source.

Anaerobic Digestion

Anaerobic digestion is a bacterial process that is carried out in the absence of oxygen. The process can either be *thermophilic* digestion, in which sludge is fermented in tanks at a temperature of 55 °C, or *mesophilic*, at a temperature of around 36 °C. Though allowing shorter retention time (and thus smaller tanks), thermophilic digestion is more expensive in terms of energy consumption for heating the sludge.

Mesophilic anaerobic digestion (MAD) is also a common method for treating sludge produced at sewage treatment plants. The sludge is fed into large tanks and held for a minimum of 12 days to allow the digestion process to perform the four stages necessary to digest the sludge. These are hydrolysis, acidogenesis, acetogenesis, and methanogenesis. In this process the complex proteins and sugars are broken down to form more simple compounds such as water, carbon dioxide, and methane.

Anaerobic digestion generates biogas with a high proportion of methane that may be used to both heat the tank and run engines or microturbines for other on-site processes. Methane generation is a key advantage of the anaerobic process. Its key disadvantage is the long time required for the process (up to 30 days) and the high capital cost. Many larger sites utilize the biogas for combined heat and power, using the cooling water from the generators to maintain the temperature of the digestion plant at the required 35 ± 3 °C. Sufficient energy can be generated in this way to produce more electricity than the machines require.

Aerobic Digestion

Aerobic digestion is a bacterial process occurring in the presence of oxygen resembling a continuation of the activated sludge process. Under aerobic conditions, bacteria rapidly consume organic matter and convert it into carbon dioxide. Once there is a lack of organic matter, bacteria die and are used as food by other bacteria. This stage of the process is known as *endogenous respiration*. Solids reduction occurs in this phase. Because the aerobic digestion occurs much faster than anaerobic digestion, the capital costs of aerobic digestion are lower. However, the operating costs are characteristically much greater for aerobic digestion because of energy used by the blowers, pumps and motors needed to add oxygen to the process. However, recent technological advances include non-electric aerated filter systems that use natural air currents for the aeration instead of electrically operated machinery.

Aerobic digestion can also be achieved by using diffuser systems or jet aerators to oxidize the sludge. Fine bubble diffusers are typically the more cost-efficient diffusion method, however, plugging is typically a problem due to sediment settling into the smaller air holes. Coarse bubble diffusers are more commonly used in activated sludge tanks or in the flocculation stages. A key component for selecting diffuser type is to ensure it will produce the required oxygen transfer rate.

Composting

Composting is an aerobic process of mixing sewage sludge with agricultural byproduct sources of carbon such as sawdust, straw or wood chips. In the presence of oxygen, bacteria digesting both the sewage sludge and the plant material generate heat to kill disease-causing microorganisms and parasites. Maintenance of aerobic conditions with 10 to 15 percent oxygen requires bulking agents allowing air to circulate through the fine sludge solids. Stiff materials like corn cobs, nut shells, shredded tree-pruning waste, or bark from lumber or paper mills better separate sludge for ventilation than softer leaves and lawn clippings. Light, biologically inert bulking agents like shredded tires may be used to provide structure where small, soft plant materials are the major source of carbon.

Uniform distribution of pathogen-killing temperatures may be aided by placing an insulating blanket of previously composted sludge over aerated composting piles. Initial moisture content of the composting mixture should be about 50 percent; but temperatures may be inadequate for pathogen reduction where wet sludge or precipitation raises compost moisture content above 60 percent. Composting mixtures may be piled on concrete pads with built-in air ducts to be covered by a layer of unmixed bulking agents. Odors may be minimized by using an aerating blower drawing vacuum through the composting pile via the underlying ducts and exhausting through a filtering pile of previously composted sludge to be replaced when moisture content reaches 70 percent. Liquid accumulating in the underdrain ducting may be returned to the sewage treatment plant; and composting pads may be roofed to provide better moisture content control.

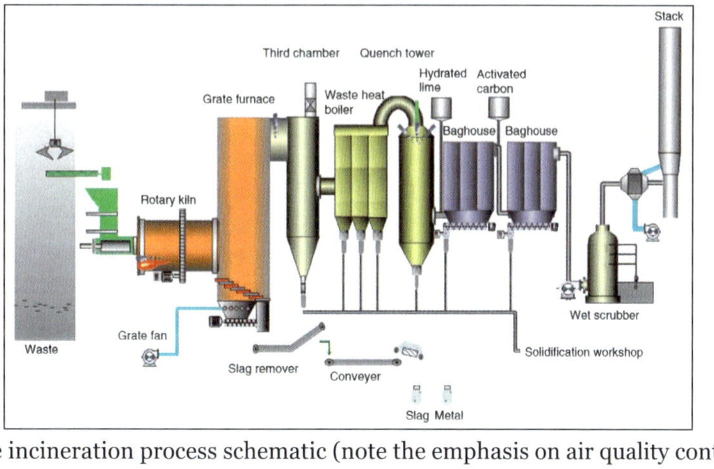

Sludge incineration process schematic (note the emphasis on air quality control).

After a composting interval sufficient for pathogen reduction, composted piles may be screened to recover undigested bulking agents for re-use; and composted solids passing through the screen may be used as a soil amendment material with similar benefits to peat. The optimum initial carbon-to-nitrogen ratio of a composting mixture is between 26-30:1; but the composting ratio of agricultural byproducts may be determined by the amount required to dilute concentrations of toxic chemicals in the sludge to acceptable levels for the intended compost use. Although toxicity is low in most agricultural byproducts, suburban grass clippings may have residual herbicide levels detrimental to some agricultural uses; and freshly composted wood byproducts may contain phytotoxins inhibiting germination of seedlings until detoxified by soil fungi.

Sewage sludge after drying in a sludge drying bed.

Incineration

Incineration of sludge is less common because of air emissions concerns and the supplemental fuel (typically natural gas or fuel oil) required to burn the low calorific value sludge and vaporize residual water. On a dry solids basis, the fuel value of sludge varies from about 9,500 British thermal units per pound (980 cal/g) of undigested sewage sludge to 2,500 British thermal units per pound (260 cal/g) of digested primary sludge. Stepped multiple hearth incinerators with high residence time and fluidized bed incinerators are the most common systems used to combust wastewater sludge. Co-firing in municipal waste-to-energy plants is occasionally done, this option being less expensive assuming the facilities already exist for solid waste and there is no need for auxiliary fuel. Incineration tends to maximize heavy metal concentrations in the remaining solid ash requiring disposal; but the option of returning wet scrubber effluent to the sewage treatment process may reduce air emissions by increasing concentrations of dissolved salts in sewage treatment plant effluent.

This simple evaporative sludge drying bed near Damascus in Syria illustrates the initial consistency of primary sludge being discharged from the primary settling tank via the pipe in the foreground.

Drying Beds

Simple sludge drying beds are used in many countries, particularly in developing countries, as they are a cheap and simple method to dry sewage sludge. Drainage water must be captured; drying beds are sometimes covered but usually left uncovered. Mechanical devices to turn over the sludge in the initial stages of the drying process are also available on the market.

Drying beds are typically composed of four layers consisting of gravel and sand. The first layer is coarse gravel that is 15 to 20 centimeters thick. Followed by fine gravel that is 10 centimeters thick. The third layer is sand that can be between 10 to 15 centimeters and serves as the filter between the sludge and gravel. Sludge dries up and water percolates to the first layer that is collected at the drainage pipe that is beneath all layers.

Emerging Technologies

The thermal hydrolysis system at the Blue Plains treatment plant in Washington, D.C. is the largest in the world as of 2016.

- Phosphorus recovery from sewage sludge or from sludge dewatering streams is receiving increased attention particularly in Sweden, Germany and Canada, as phosphorus is a limited resource (a concept also known as "peak phosphorus") and is needed as fertilizer to feed a growing world population. Phosphorus recovery methods from wastewater or sludge can be categorized by the origin of the used matter (wastewater, sludge liquor, digested or non-digested sludge, ash) or by the type of recovery processes (precipitation, wet-chemical extraction and precipitation, thermal treatment). Research on phosphorus recovery methods from sewage sludge has been carried out in Sweden and Germany since around 2003, but the technologies currently under development are not yet cost effective, given the current price of phosphorus on the world market.
- The Omni Processor is a process that is currently under development that treats sewage sludge and can generate a surplus of electrical energy if the input materials have the right level of dryness.

- Thermal depolymerization produces light hydrocarbons from sludge heated to 250 °C and compressed to 40 MPa.
- Thermal hydrolysis is a two-stage process combining high-pressure boiling of sludge, followed by a rapid decompression. This combined action sterilizes the sludge and makes it more biodegradable, which improves digestion performance. Sterilization destroys pathogens in the sludge resulting in it exceeding the stringent requirements for land application (agriculture). Thermal hydrolysis systems are operating at sewage treatment plants in Europe, China and North America, and can generate electricity as well as high quality sludge.

Disposal or use as Fertilizer

When a liquid sludge is produced, further treatment may be required to make it suitable for final disposal. Sludges are typically thickened and/or dewatered to reduce the volumes transported off-site for disposal. Processes for reducing water content include lagooning in drying beds to produce a cake that can be applied to land or incinerated; pressing, where sludge is mechanically filtered, often through cloth screens to produce a firm cake; and centrifugation where the sludge is thickened by centrifugally separating the solid and liquid. Sludges can be disposed of by liquid injection to land or by disposal in a landfill.

There is no process which completely eliminates the need to dispose of treated sewage sludge.

Much sludge originating from commercial or industrial areas is contaminated with toxic materials that are released into the sewers from the industrial processes. Elevated concentrations of such materials may make the sludge unsuitable for agricultural use and it may then have to be incinerated or disposed of to landfill.

Edmonton, Alberta and Canada

The Edmonton Composting Facility, in Edmonton, Alberta, Canada, is the largest sewage sludge composting site in North America.

New York City, U.S.

Sewage sludge can be superheated and converted it into pelletized granules that are high in nitrogen and other organic materials. In New York City, for example, several sewage treatment plants have dewatering facilities that use large centrifuges along with the addition of chemicals such as polymer to further remove liquid from the sludge. The product which is left is called "cake," and that is picked up by companies which turn it into fertilizer pellets. This product, also called biosolid, is then sold to local farmers and turf farms as a soil amendment or fertilizer, reducing the amount of space required to dispose of sludge in landfills.

Southern California, U.S.

In the very large metropolitan areas of southern California inland communities return sewage sludge to the sewer system of communities at lower elevations to be reprocessed at a few very large treatment plants on the Pacific coast. This reduces the required size of interceptor sewers and allows local recycling of treated wastewater while retaining the economy of a single sludge processing facility and is an example of how sewage sludge can help solve an energy crisis.

ORGANIC WASTE MANAGEMENT

Organic Waste Management (OWM) operates an organic waste removal service that collects "source-separated" food scraps from your premises and disposes them off in an environmentally sustainable method.

Composting is a managed process which utilizes microorganisms naturally present in organic matter and soil to decompose organic material. These microorganisms require basic nutrients, oxygen, and water in order for decomposition to occur at an accelerated pace. The end-product, compost, is a dark brown, humus-like material which can be easily and safely handled, stored, and used as a valuable soil conditioner. The composting process is dependent upon several factors, including: the population of microorganisms, carbon to nitrogen ratio, oxygen level, temperature, moisture, surface area, pH, and time.

Aerobic Composting

The composting process involves microorganisms feeding on organic material and consuming oxygen.The process generates heat, drives off moisture, and reduces bulky organic waste into a beneficial soil-like material containing nutrients, humus and microorganisms in just a few months. Material in an unmanaged pile of organic debris will eventually break down but the process will take a long time and may result in odor or other nuisance problems due to poor aeration.

What to Compost?

Most organic waste materials generated by a greenhouse can be composted. Large material will need to be shredded before it is added to a carefully-constructed compost pile. Some material may begin to decompose in a storage pile but full composting will not occur until the material is mixed and managed in the correct proportions of carbon to nitrogen (C:N ratio), with adequate airflow and moisture.

Composting is an excellent method of recycling grass clippings. However, do not compost grass clippings or any other plant residues that have been treated with herbicides.If carried out properly, it can reduce the potential weed seeds and diseases from being reintroduced into the fields.The finished compost is a stable organic material which is a useful soil conditioner or nutrient source. Due to the characteristics of fresh grass clippings (high-moisture, high-nitrogen content and small particle size), co-composting with a high-carbon bulking agent is essential.

Unacceptable materials for composting: Chemically treated wood products, plastic (e.g. pots, bags and sheet film), unprocessed sod and chunks of soil, large bulky idems (e.g. stumps, pallets, concrete and asphalt).

Acceptable Materials for composting: Green and woody plant clippings and trimmings, soil, plant media, untreated wood and uncoated paper scraps.

Site Selection for Compost Piles

Proper site selection is a prerequisite to the establishment of safe and effective composting operations. The location of a composting operation directly impacts the amount of site preparation required and the measures needed to satisfy environmental and regulatory requirements.

The Basic Composting Process

The general steps in the biological process which creates compost are the same regardless of the raw materials being composted or the size and complexity of the production facility. A compost must pass through all of the steps outlined here in order for it to be considered of high enough quality for use in organic potting mixes.

The progress of organic matter decomposition during composting can be followed by monitoring the temperature of the compost pile. During the initial phase of composting the temperature of the pile increases rapidly as the population and activity of decay microorganisms increases in response to the readily decomposable carbon in the raw materials. The goals are to reach a temperature between 131°F or more and to maintain this temperature range until the microorganisms begin to exhaust the readily available carbon. During composting the pile is turned and remixed several times to ensure complete heating and decomposition.

To comply with the National Organic Program standards compost piles must maintain 131-170 °F for at least 3 days (static pile) or at least 15 days (windrow, turned at least 5 times). High temperatures are necessary to kill any human pathogens especially if farm manure is a component. Also, weed seeds and plant diseases are most successfully killed at high temperatures. Most weed seeds are destroyed at 145 °F.

Following the high temperature phase there is an extended period of gradual temperature decline until the pile reaches ambient air temperature. Now, if the pile is turned, reheating will not occur. At this point the compost is said to be "near maturity", but to ensure that the compost is stable and ready to use, most producers allow some extra time for the compost to "cure". How long composting lasts varies with the method. It could take about 1-2 years in a static unturned pile, 6-9 months if the pile is turned occasionally, or only 1-4 months the pile is turned frequently.

Many types of raw materials can be used for making compost; some common materials are listed in the following table. Pay close attention to the comments in the table.

Compost Material	Comments
Farm animal manure	Must be composted
Straw and bedding	
Crop residues	Must be pesticide free
Fruit & vegetable wastes	Must be pesticide free
Food processing wastes	
Seafood processing waste	
Grass clippings	Must be pesticide free
Sawdust & other wood wastes	Use in moderation, low nutrient value.
Newspaper	Black ink only, shredded, <25%
Leaves	Shredded

It is important that the raw materials be properly prepared prior to mixing and the start of composting. Most organic materials must be shredded or ground to reduce particle size and help make them less resistant to decay.

During composting, oxygen and moisture levels are critical factors in determining the degree of decomposition which takes place and the length of time it takes to reach a stable product. Oxygen levels below 5% and moisture levels above or below the range of 40-65% inhibit the composting process. Most composting operations aerate the piles (turn) and irrigate them if conditions favor excessive drying. Too little moisture will inhibit microbial activity and slow down the composting process, while too much moisture will restrict the flow of oxygen because all pore space is taken up by water instead of air, and anaerobic conditions will begin to develop. The volume of the finished compost is smaller than the volume of raw materials because of the breakdown of organic matter and the evaporation of water.

When is the compost ready for use? Currently there is no single widely accepted criterion to determine when compost is "done". Measurements of temperature, respiration, ammonia production, pH, and carbon to nitrogen ratio (C:N) are among the potential indicators of compost maturity, but no one factor is completely reliable. Generally, at the end of active composting (heating period) producers allow a "curing" period of about 1-2 months to make sure the compost is stable before it's used.

- Most organic waste materials generated by a greenhouse can be composted.
- Avoid composting grass clippings that has been treated with herbicides.
- Compost piles should always be distant and downwind from sensitive neighbors and not sited close to residential property.
- Piles should be protected from surface water and storm water runoff.
- Proportions of carbon to nitrogen are critical to successful composting. The materials being composted will determine the exact recipe for any given operation. Materials with high carbon to nitrogen ratios, such as 100:1, should be balanced with materials having low carbon to nitrogen ratios, e.g. 15:1.
- Regular turning of the pile will mix the nutrients and re-establish pile structure.
- Moisture content can be adjusted during turning.
- After the compost has gone through several heating and cooling cycles and the original waste has decomposed, the compost process should slowly finish in a curing pile.
- Activities that tend to release odors should be scheduled to minimize negative impacts.
- Consider wind conditions before opening compost piles. Stronger winds can disperse odors but also create dust concerns. Care should be taken to control dust when grinding and turning piles.
- Most odor problems can be avoided, controlled, or minimized by keeping the compost pile aerobic, porous, well aerated, and well mixed. Odor problems are most likely when anaerobic decomposition is occurring.

References

- Solid-waste-management, waste-management, environment, energy: vikaspedia.in, Retrieved 15 April, 2019
- Biological-sewage-treatment-plant: marineengineeringonline.com, Retrieved 1 June, 2019
- Wastewater engineering: treatment and reuse (4th ed.). Metcalf & eddy, inc., mcgraw hill, usa. 2003. P. 1449. Isbn 0-07-112250-8
- Food-waste-disposal: ecomena.org, Retrieved 19 May, 2019
- Sewage-treatment-physical-and-biological-treatment-of-sewage,biology: yourarticlelibrary.com, Retrieved 1 August, 2019
- Sforza, teri. "new plan replaces sewage sludge fiasco". Orange county register. Retrieved 15 january 2015
- Organic-waste-management, greenhouse-best-management-practices-bmp-manual, greenhouse-floriculture: umass.edu, Retrieved 10 May, 2019

4

Processes for Biodegradable Waste Management

Some of the major concepts and processes related to biodegradable waste management are decomposition, aerobic digestion, anaerobic digestion, pyrolysis, bioremediation, biodegradation, etc. The chapter closely examines these key concepts and processes of biodegradable waste management.

DECOMPOSITION

Decomposition is the natural process of dead animal or plant tissue rotting or being broken down. This process is carried out by invertebrates, fungi and bacteria. The result of decomposition is that the building blocks required for life can be recycled. The process of decomposition is a gradual complex process, one in which both chemical and biological processes must occur in order for organic matter to change into compost.

The decomposition (stabilization) of organic matter by biological action has been taking place in nature since life first appeared on our planet. In recent times, man has attempted to control and directly utilize the process for sanitary recycling and reclamation of organic waste material. Such organic materials as vegetable matter, animal manure and other organic refuse can be converted from otherwise wasted materials to a more stable form for use as a soil amendment by this process. This process is called "composting" and the final product of composting is called "compost". Generally speaking there are two processes that yield compost:

- Anaerobic (without oxygen) decomposition.
- Aerobic (with oxygen) decomposition and stabilization.

In these processes, bacteria, fungi, molds, protozoa, actinomycetes, and other saprophytic organisms feed upon decaying organic materials initially, while in the later stages of decomposition mites, millipedes, centipedes, springtails, beetles and earthworms further breakdown and enrich the composting materials. The organisms will vary in the pile due to temperature conditions, but the goal in composting is to create the most favorable environment possible for the desired organisms.

Differences between aerobic and anaerobic composting are discussed below:

Anaerobic Decomposition (Fermentation)

Anaerobic decomposition takes place in nature, as in the decomposition of the organic muds at the bottom of marshes and in buried organic materials to which oxygen does not have access. Intensive reduction of organic matter by putrefaction is usually accompanied by disagreeable odors of hydrogen sulfide and reduced organic compounds which contain sulfur, such as mercaptans (any sulfur-containing organic compound).

Putrefactive breakdown of organic material takes place anaerobically. Organic compounds break down by the action of living organisms that do not require air in the normal sense. These organisms use nitrogen, phosphorus, and other nutrients to live and to develop cell protoplasm, but they reduce the organic nitrogen to organic acids and ammonia. The carbon from the organic compounds which is not utilized in the cell protein is liberated mainly in the reduced form of methane (CH_4). A small portion of carbon may be respired as carbon dioxide (Co_2).

Since anaerobic destruction of organic matter is a reduction process, the final product, humus, is subject to some aerobic oxidation when put on the soil, that is, it may appear to decompose further after being exposed to air. This oxidation is minor, takes place rapidly, and is of no consequence in the utilization of the material on the soil. In other words, much less heat is generated in anaerobic decomposition than in aerobic decomposition.

The lack of heat generated in the anaerobic destruction of organic matter is a definite disadvantage if contaminated materials areused for composting. High temperatures are needed for the destruction of pathogens and parasites. In anaerobic decomposition the pathogenic organisms do eventually disappear in the organic mass, as a result of the unfavorable environment and biological antagonisms. The disappearance is slow, and the material must be held for periods of six months to a year to ensure relatively complete destruction of pathogens, such as the eggs of Ascaris, nematodes which are among the most resistant of the fecal-borne disease parasites in wastes. Therefore, make compost this year and use it next year.

However, organic material can be decomposed anaerobically to produce compost. For instance, a heavy plastic bag can be used to decompose grass clippings or other high nitrogen materials, shredded leaves, kitchen trimmings, a small amount of stable manure or other compostable materials. However, as anaerobic compost can have a strong odor (and may need to be aired prior to using), it is not usually the first choice for home owners.

Aerobic Decomposition

When organic materials decompose in the presence of oxygen, the process is called "aerobic." The aerobic process is most common in nature. For example, it takes place on ground surfaces such as the forest floor, where droppings from trees and animals are converted into a relatively stable humus. There is no accompanying bad smell when there is adequate oxygen present.

In aerobic decomposition, living organisms, which use oxygen, feed upon the organic matter. They use the nitrogen, phosphorus, some of the carbon, and other required nutrients. Much of the carbon serves as a source of energy for the organisms and is burned up and respired as carbon dioxide

(Co_2). Since carbon serves both as a source of energy and as an element in the cell protoplasm, much more carbon than nitrogen is needed. Generally about two-thirds of carbon is respired as Co_2, while the other third is combined with nitrogen in the living cells. However, if the excess of carbon over nitrogen (C:N ratio) in organic materials being decomposed is too great, biological activity diminishes. Several cycles of organisms are then required to burn most of the carbon.

When some of the organisms die, their stored nitrogen and carbon becomes available to other organisms. As other organisms use the nitrogen from the dead cells to form new cell material, once more excess carbon is converted to Co_2. Thus, the amount of carbon is reduced and the limited amount of nitrogen is recycled. Finally, when the ratio of available carbon to available nitrogen is in sufficient balance, nitrogen is released as ammonia. Under favorable conditions, some ammonia may oxidize to nitrate. Phosphorus, potash, and various micro-nutrients are also essential for biological growth. These are normally present in more than adequate amounts in compostable materials and present no problem.

During composting a great deal of energy is released in the form of heat in the oxidation of the carbon to Co_2. For example, if a gram-molecule of glucose is dissimilated under aerobic conditions, 484 to 674 kilogram calories (kcal) of heat may be released. If the organic material is in a pile or is otherwise arranged to provide some insulation, the temperature of the material during decomposition will rise to over 170 °F. If the temperature exceeds 162 °F to 172 °F, however, the bacterial activity is decreased and stabilization is slowed down.

Initially, mesophilic organisms, which live in temperatures of 50 °F to 115 °F, colonize in the materials. When the temperature exceeds about 120 °F, thermophilic organisms, which grow and thrive in the temperature range 115 °F to 160 °F., develop and replace the mesophilic bacteria in the decomposition material. Only a few groups of thermophiles carry on any activity above 160 °F.

Oxidation at thermophilic temperatures takesplace more rapidly than at mesophilic temperatures and, hence, a shorter time is required for decomposition (stabilization). The high temperatures will destroy pathogenic bacteria, protozoa (microscopic one-celled animals), and weed seeds, which are detrimental to health or agriculture when the final compost is used.

Aerobic oxidation of organic matter produces no objectionable odor. If odors are noticeable, either the process is not entirely aerobic or there are some special conditions or materials present which are creating an odor. Aerobic decomposition or composting can be accomplished in pits, bins, stacks, or piles, if adequate oxygen is provided. Turning the material at intervals or other techniques for adding oxygen is useful in maintaining aerobic conditions.

Compost piles under aerobic conditions attain a temperature of 140 °F to 160 °F in one to five days depending upon the material and the condition of the composting operation. This temperature can also be maintained for several days before further aeration. The heat necessary to produce and maintain this temperature must come from aerobic decomposition which requires oxygen. After a period of time, the material will become anaerobic unless it is aerated.

Here the term "aerobic composting" will be used in its commonly accepted meaning of that process. It requires a considerable amount of oxygen and produces none of the characteristic features of anaerobic putrefaction. In its modern sense, aerobic composting can be defined as a process in which, under suitable environmental conditions, aerobic organisms, principally thermophilic, utilize considerable amounts of oxygen in decomposing organic matter to a fairly stable humus.

Requirements for Efficient Decomposition

Decomposition of organic material in the compost pile depends on maintaining microbial activity. Any factor which slows or halts microbial growth also impedes the composting process. Efficient decomposition occurs if aeration, moisture, particle size, and a sufficient source of carbon and nitrogen are in evidence.

Aeration/Oxygen

Oxygen is required for microbes to decompose organic wastes efficiently. Some decomposition occurs in the absence of oxygen (anaerobic conditions); however, the process is slow, and foul odors may develop. Because of the odor problem, composting without oxygen is not recommended in a residential setting unless the process is conducted in a fully closed system. Mixing the pile once or twice a month provides the necessary oxygen and significantly hastens the composting process. A pile that is not mixed may take three to four times longer to decompose. Raising the pile off the ground allows air to be drawn through the mass as the material decomposes. Coarse materials should be placed on the bottom as the pile is built or placed in the pile and removed after the decomposition starts.

Oxygen levels should be kept at 5% throughout the entire pile. Typical oxygen percents range from 6% – 16% in the pile air spaces or in the exhausted air; and 20% at the exposed portions of the pile. Failure to keep all parts of the compost pile above the 5% oxygen level will cause the pile to "go anaerobic", with the accompanying odor problems. The more oxygen, up to at least 10-12 percent, the more quickly the biodegradation will take place.

Moisture

Adequate moisture is essential for microbial activity. A dry compost will not decompose efficiently. Proper moisture encourages the growth of microorganisms that break down the organic matter into humus. If rainfall is limited, water the pile periodically to maintain a steady decomposition rate. Add enoughwater so the pile is damp but not soggy. Avoid over watering. Excess water can lead to anaerobic conditions which slow down the degradation process and cause foul odors. If the pile should become too wet, turn it to dry it out and restart the process.

Particle Size

Grinding the organic material before composting greatly reduces decomposition time. The smaller the size of the organic refuse particle, the more quickly it can be consumed by the microbes. A shredder is useful for chipping or shredding most landscape refuse and is essential if brush or sticks are to be composted. A low-cost method of reducing the size of fallen tree leaves is to mow the lawn before raking. Wind-rowing the leaves into long narrow piles one foot high will make the shredding process more efficient. If the mower has an appropriate bag attachment, the shredded leaves can be collected directly. However, grinding is entirely optional.

Temperature

Temperature of the compost pile is very important to the biological activity taking place. Low outside temperatures slow the activity down, while warmer temperatures speed up decomposition. The microbes that make up the bulk of the decomposition process fall into two categories: mesophilic, those that live and function in temperatures of 50 to 113 °F, and thermophilic, those that thrive at temperatures between 113 to 158 °F. A well-mixed, adequately working compost pile will heat to temperatures between 110 °F and 160 °F as the microbes actively feed on the organic materials. These high temperatures will help destroy weed seeds and disease organisms within the pile.

Organisms in Composting

There are many organisms that breakdown organic materials. Most are not seen by the human eye, but they are there throughout the process. Others that are large enough to see, are usually associated with the later breakdown stages.

The most important organisms in the breakdown process are the bacteria. The bacteria present in any given pile are dependent upon the raw material present, amount of air in the pile, moisture conditions of the pile, pile temperature and numerous other factors. Bacteria are very small–20,000+ laid end to end may span no more that an inch. Compostable organic materials normally contain a large number and many different types of bacteria, fungi, molds, and other living organisms. Only very limited data are available regarding the variety of different organisms and their specific functions. It appears that more species of bacteria are involved in aerobic decomposition than in anaerobic putrefaction.

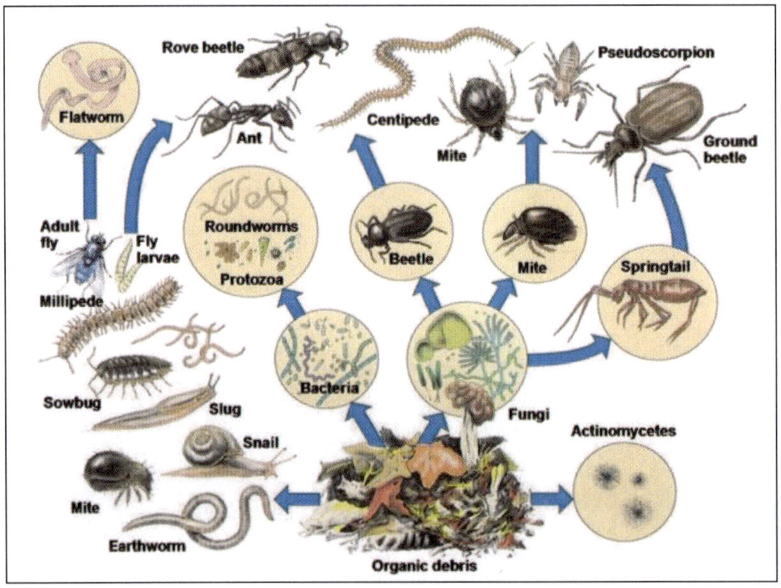

Although many types of organisms are required for decomposition of the different materials, the necessary variety is usually present in the materials to be composted, and the organisms thrive when environmental conditions are satisfactory. During decomposition, marked changes take place in the nature and abundance of the biological population. Some of the many species will multiply rapidly at first but will dwindle as the environment changes and other organisms are able to thrive under more varied conditions. Temperature and changes in the available food supply probably exert the greatest influence in determining the species of organisms comprising the population at any one time. Aerobic composting is a dynamic process in which the work is done by the combined activities of a wide succession of mixed bacteria, actinomycetes, fungi, and other biological populations. Since each is suited to a particular environment of relatively limited duration and each is most active in decomposition of some particular type of organic matter, the activities of one group complement those of another. The mixed populations parallel the complex environments afforded by the heterogeneous nature of the compostable material. Except for short periods during turning, the temperature increases steadily in proportion to the amount of biological activity until equilibrium (state of balance) with subsequent heat losses is reached, or the material becomes well-stabilized (humus-like).

In aerobic composting bacteria, actinomycetes, and fungi are the most active. Mesophilic (low temperature 50-115 °F) bacteria are characteristically predominant at the start and in the early part of the process, soon giving way to thermophilic (high temperature 110-150+ °F) bacteria which inhabit all parts of the stack where the temperature is satisfactory, eventually most of the stack. Thermophilic fungi usually appear after 5 to 10 days, and actinomycetes become prominent in the final stages, when short duration, rapid composting is accomplished. Except in the final stages of the composting period, when the temperature drops, actinomycetes and fungi are confined to a sharply defined outer zone of the stack, 2 to 6 inches in thickness, beginning just under the outer surface. Some molds also grow in this outer zone. The population of fungi and actinomycetes is often great enough to impart a distinctly grayish white appearance to this outer zone.

The sharply defined inner and outer limits of the shell (in which actinomycetes and fungi grow during the high temperature active composting period) are due to the inability of these organisms to grow at the higher temperatures of the interior of the pile. The thermophilic actinomycetes and fungi have been found to grow at temperatures between 120° and 150 °F. Frequent turning–such as is sometimes necessary for fly control–inhibits their growth, since the cooler outer shell is turned into the interior before they can develop in large numbers.

Various investigations have shown that many different types of thermophilic bacteria apparently play a major part in decomposing protein and other organic matter. They appear to be solely responsible for the intense activity characteristic of the first few days, when the temperature reaches 150° to 170 °F and major changes in the nature of the compost pile are taking place, i.e., when the pile is drastically shrinking and the appearance of the material is undergoing rapid change. They continue to predominate throughout the process in theinterior of the piles, where temperatures are inhibitory to actinomycetes and fungi.

In spite of being confined primarily to the outer layers and becoming active only during the latter part of the composting period, fungi and actinomycetes play an important role in the decomposition of cellulose, lignins, and other more resistant materials, which are attacked after the more readily decomposed materials have been utilized. There are many bacteria which attack cellulose. However, in the parts of compost piles populated chiefly by bacteria, cellulose (paper) breaks down very little, whereas in the layers or areas inhabited by actinomycetes and fungi it becomes almost unrecognizable. Considerable cellulose and lignin decomposition by actinomycetes and fungi can occur near the end of the composting period when the temperatures have begun to drop and the environment in a larger part of the pile is satisfactory for their growth.

It should be noted that since the necessary organisms for composting are usually present and will carry on the process when the environment is suitable, an extensive knowledge of the characteristics of the various organisms is not necessary for understanding a compost pile. Normal maintenance as described in this manual will help to insure proper balance and numbers of beneficial microorganisms.

Since decomposition is the crux of the composting process, let's take a look at the various organisms that play an essential role in the working compost heap. Most are microscopic, some are large enough to be observed with the unaided eye, but all are beneficial, each having a role in breaking down raw organic matter into finished compost. They are known as decomposers. By far the most important microscopic decomposers are bacteria, which do the lion's share of decomposition in the compost heap. But there are other microscopic creatures such as actinomycetes, fungi, and protozoa, that also play an important role. Together, these are chemical decomposers that change the chemistry of the organic wastes. The largerfauna in the heap include mites, millipedes, flatworms, centipedes, sowbugs, snails, slugs, spiders, springtails, beetles, ants, flies, nematodes and, most importantly, earthworms. Collectively, these are called the physical decomposers since they bite, grind, suck, tear and chew the materials into smaller pieces, making them more suitable for the chemical work of the microscopic decomposers.

All of the organisms, from the microscopic bacteria to the largest of the physical decomposers, are part of a complex food chain in the compost pile. They can be categorized as first, second and third level consumers, depending upon whom they eat and by whom they are eaten. First level consumers attract and become the food of second level consumers, who in turn are consumed by third level consumers. The organisms comprising each level of the food chain serve to keep the populations of the next lower level in check, so that a balance can be maintained throughout the compost. For example, according to Daniel L. Dindal, in Ecology of Compost:

> "Mites and springtails eat fungi. Tiny feather-winged beetles feed on fungal spores. Nematodes ingest bacteria. Protozoa and rotifers present in water films feed on bacteria and plant particles. Predaceous mites and pseudo- scorpions prey upon nematodes, fly larvae, other mites and collembolans. Free-living flatworms ingest gastropods, earthworms, nematodes and rotifers. Third level consumers such as centipedes, rove beetles, ground beetles, and ants prey on second level consumers. These creatures function best at medium or mesophilic temperatures, so they will not be in the pile at all times."

Chemical Decomposers

These organisms are the initial inhabitants of the pile. Many of them are unseen and come in with the materials that make up the pile. These organisms are around all of the time and only need to find the conditions right "to their liking" in order to start their normal functions of breaking down organic materials.

Bacteria

Bacteria likely to be found in a compost heap are aerobic bacteria that specialize in breaking down organic compounds and thrive in temperatures ranging up to 170 °F (77 °C). Bacterial populations differ from pile to pile, depending upon the raw materials of the compost, degree of heat, amount of air present, moisture level, geographical location of the pile, and other considerations. Bacteria are single-celled and can be shaped like a sphere, rod, or a spiral twist. They are so small that it would take 25,000 bacteria laid end to end to take up one inch on a ruler, and an amount of garden soil the size of a pea may contain up to a billion bacteria. Most bacteria are colorless and cannot make carbohydrates from sunshine, water, and carbon dioxide the way more complex green plants can. Some bacteria produce colonies; others are free-living. All reproduce by means of binary fission.

In binary fission, the nucleus splits in two and a new cell wall grows crosswise over the middle of the cell. Each half contains one of the two nuclei, so that a new individual is produced from a single bacterial cell.

Under the best conditions, a colony of bacteria can multiply into billions in a very short time. The life span of one generation of bacteria is about 20 to 30 minutes, so that one cell may yield a progeny of billions of individuals in half a day.

Bacteria are the most nutritionally diverse of all organisms, which is to say, as a group, they can eat nearly anything. Most compost bacteria are heterotrophic, meaning that they can use living or dead organic materials. Some are so adaptable that they can use more than a hundred different organic compounds as their source of carbon because of their ability to produce a variety of enzymes. Usually, they can produce the appropriate enzyme to digest whatever material they find themselves on. In addition, respiratory enzymes in the cell membrane make aerobic respiration possible as an energy source for compost bacteria.

Since bacteria are smaller, less mobile and less complex than most organisms, they are less able to escape an environment that becomes unfavorable. A decrease in the temperature of the pile or a sharp change in its acidity can render bacteria inactive or kill them. When the environment of a heap begins to change, bacteria that formerly dominated may be decimated by another species.

Actinomycetes

The characteristically earthy smell of newly plowed soil in the spring is caused by actinomycetes, a higher form of bacteria similar to fungi and molds. Actinomycetes are especially important in the formation of humus. While most bacteria are found in the top foot or so of topsoil, actinomycetes may work many feet below the surface. Deep under the roots they convert dead plant matter to a peat-like substance. While they are decomposing animal and vegetable matter, actinomycetes liberate carbon, nitrogen and ammonia, making nutrients available for higher plants. They are found on every natural substrate, and the majority are aerobic and mesophilic. Five percent or more of the soil's bacterial population is comprised of actinomycetes.

The reason bacteria tend to die rapidly as actinomycete populations grow in the compost pile is that actinomycetes have the ability to produce antibiotics, chemical substances that inhibit bacterial growth.

Protozoa

Protozoa are the simplest form of animal organism. Even though they are single-celled and microscopic in size, they are larger and more complex in their activities than most bacteria. A gram of soil can contain as many as a million protozoa, but a gram of compost has many thousands less, especially during the thermophilic stage. Protozoa obtain their food from organic matter in the same way bacteriado, but because they are present in far fewer numbers than are bacteria, they play a much smaller part in the composting process.

Fungi

Fungi are many-celled, filamentous or single-celled primitive plants. Unlike more complex green plants, they lack chlorophyll, and, therefore, lack the ability to make their own carbohydrates. Most of them are classified as saprophytes because they live on dead or dying material and obtain energy by breaking down organic matter in dead plants and animals. Like the actinomycetes, fungi take over during the final stages of the pile when the compost has been changed to a more easily digested form. The best temperature for active fungi in the compost heap is around 70° to 75 °F though some thermophilic forms prefer much greater heat and survive to 120 °F.

Physical Decomposers

The larger organisms that chew and grind their way through the compost heap are higher up in the food chain and are known as physical decomposers. The following is a rundown of some of the larger physical decomposers that you may find in nearly any compost heap. Most of these creatures function best at medium or mesophilic temperatures, so they will not be in the pile at all times.

Mites

Mites are related to ticks, spiders, and horseshoe crabs because they have in common six leg-like, jointed appendages. They can be free-living or parasitic, sometimes both at once. Some mites are small enough to be invisible to the naked eye, while some tropical species are up to a half-inch in length. Mites reproduce very rapidly, moving through larval, nymph, adult and dormant stages. They attack plant matter, but some are also second level consumers, ingesting nematodes, fly larvae, other mites and springtails.

Millipedes

The wormlike body of the millipede has many leg-bearing segments, each exceptthe front few bearing two pairs of walking legs. The life cycles are not well understood, except that eggs are laid in the soil in springtime, hatching into small worms. Young millipedes molt several times before gaining their full complement of legs. When they reach maturity, adult millipedes can grow to a length of 1 to 2 inches. They help break down plant material by feeding directly on it.

Centipedes

Centipedes are flattened, segmented worms with 15 or more pairs of legs, 1 pair per segment. They hatch from eggs laid during the warm months and gradually grow to their adult size. Centipedes are third level consumers, feeding only on living animals, especially insects and spiders.

Sowbugs

The sowbug is a fat-bodied, flat creature with distinct segments. In structure, it resembles the crayfish to which it is related. Sowbugs reproduce by means of eggs that hatch into smaller versions of the adults. Since females are able to deposit a number of eggs at one time, sowbugs may become abundant in a compost heap. They are first level consumers, eating decaying vegetation.

Snails and Slugs

Both snails and slugs are mollusks and have muscular disks on their undersides that are adapted for a creeping movement. Snails have a spirally curved shell, a broad retractable foot, and a distinct head. Slugs, on the other hand, are so undifferentiated in appearance that one species is frequently mistaken for half of a potato. Both snails and slugs lay eggs in capsules or gelatinous masses and progress through larval stages to adulthood. Their food is generally living plant material, but they will attack fresh garbage and plant debris and will appear in the compost pile. It is well,therefore, to look for them when you spread your compost, for if they move into your garden, they can do damage to crops.

Spiders

Spiders, which are related to mites, are one of the least appreciated animals in the garden. These eight-legged creatures are third level consumers that feed on insects and small invertebrates, and they can help control garden pests.

Springtails

Springtails are very small insects, rarely exceeding one-quarter inch in length. They vary in color from white to blue-grey or metallic and are mostly distinguished by their ability to jump when disturbed. They feed by chewing decomposing plants, pollen, grains, and fungi.

Beetles

The rove beetle, ground beetle, and feather-winged beetle are the most common beetles in compost. Feather-winged beetles feed on fungal spores, while the larger rove and ground beetles prey on other insects as third level consumers. Beetles are easily visible insects with two pairs of wings, the more forward-placed of these serving as a cover or shield for the folded and thinner back-set ones that are used for flying. A beetle's immature stage is as a soft-skinned grub that feeds and grows during the warm months. Once grubs are full grown, they pass through a resting or pupal stage and change into hard-bodied, winged adults. Most adult beetles, like the larval grubs of their species, feed on decaying vegetables, while some, like the rove and ground beetles, prey on snails, insects, and other small animals. The black rove beetle is an acknowledged predator of snails and slugs. Some people import them to their gardens when slugs become a garden problem.

Ants

Ants feed on a variety of material, including aphid honeydew, fungi, seeds, sweets, scraps, other insects, and sometimes other ants. Compost provides some of these foods, and it also provides shelter for nests and hills. They will remain, however, only while the pile is relatively cool. Ants prey on first level consumers, and may benefit the composting process by bringing fungi and other organisms into their nests. The work of ants can make compost richer in phosphorus and potassium by moving minerals from one place to another.

Flies

Many flies, including black fungus gnats, soldier flies, minute flies, and houseflies, spend their larval phase in compost as maggots. Adults can feed upon almost any kind of organic material. All flies undergo egg, larval, pupal, and adult stages. The eggs are laid in various forms of organic matter. Houseflies are such effective distributors of bacteria that when an individual fly crawls across a sterile plate of lab gelatin, colonies of bacteria later appear in its tracks. You can see how during the early phases of the composting process, flies provide ideal airborne transportation for bacteria on their way to the pile. If you keep a layer of dry leaves or grass clippings on top of your pile and cover your garbage promptly while building compost, your pile will not provide a breeding place for horseflies, mosquitoes, or houseflies which may become a nuisance to humans. Fly larvae will not survive the thermophilic temperatures in the well-managed compost pile. Mites and other organisms in the pile also keep fly larvae reduced in number. However, though many flies die with the coming of frost, the rate of reproduction is so rapid that a few survivors can repopulate an area before the warm season has progressed very far.

Worms

Nematodes or eelworms, free-living flatworms, and rotifers all can be found in compost. Nematodes are microscopic creatures that can be classified into three categories: those that live on decaying organic matter; those that arepredators on other nematodes, bacteria, algae, protozoa, etc.; and those that can be serious pests in gardens where they attack the roots of plants. Flatworms, as their name implies, are flattened organisms that are usually quite small in their free-living form. Most flatworms are carnivorous and live in films of water within the compost structure. Rotifers are small, multicellular animals that live freely or in tubes attached to a substrate in the pile. Their bodies are round and divisible into three parts, a head, trunk, and tail. They are generally found in films of water and many forms are aquatic. The rotifers in compost are found in water which adheres to plant substances where they feed on microorganisms.

Earthworms

If bacteria are the champion microscopic decomposers, then the heavyweight champion is doubtlessly the earthworm. Pages of praise have been written to the earthworm, ever since it became known that this creature spends most of its time tilling and enriching the soil. The great English naturalist, Charles Darwin, was the first to suggest that all the fertile areas of this planet have at least once passed through the bodies of earthworms.

The earthworm consists mainly of an alimentary canal which ingests, decomposes, and deposits casts continually during the earthworm's active periods. As soil or organic matter is passed through an earthworm's digestive system, it is broken up and neutralized by secretions of calcium carbonate from calciferous glands near the worm's gizzard. Once in the gizzard, material is finely ground prior to digestion. Digestive intestinal juices rich in hormones, enzymes, and other fermenting substances continue the breakdown process. The matter passes out of the worm's body in the form of casts, which are the richest and finest quality of all humus material. Fresh casts are markedly higher in bacteria, organic material, available nitrogen, calcium and magnesium, and available phosphorus and potassium than soil itself.

Earthworms thrive on compost and contributeto its quality through both physical and chemical processes, and reproduce readily in the well-managed pile. Since earthworms are willing and able to take on such a large part in compost making, it is the wise gardener who adjust his composting methods to take full advantage of the earthworm's special talents.

Use of Inocula

Composting developments have been accompanied by considerable discussion of the importance of special inocula (bacterial activators), supposedly containing several pure strains of laboratory organisms or other biological factors essential in the decomposition of organic matter and nitrogen fixation, e.g., "enzymes," "hormones," "preserved living organisms," "activated factors," "biocatalyst," etc. In fact, several commercial composting processes are built around the use of some special inoculum, often known only to its discoverer and proponent, who claims it to be fundamental to the successful operation of the process. The need of such inocula has always been debatable, and most composting studies have strongly indicated that they are unnecessary.

That inocula in composting organic waste containing refuse, manure, vegetable wastes, etc. are not necessary or advantageous seems logical, since bacteria are always present in very large numbers in such material and can be eliminated only by drastic sterilization methods. In any case, the number of bacteria is rarely a limiting factor in composting. Provided that the environmental factors are appropriate, the indigenous bacteria, which are much better adapted than forms produced under laboratory conditions, multiply rapidly. Thus the rate of composting is governed simply by the environmental conditions.

Special enzyme preparations are also advertised as necessary compost "starters." However, the vast number of enzymes involved in decomposition, as well as the difficulty and expense involved in isolating and synthesizing them, would make initiating the composting process with enzymes alone highly impracticaleven if satisfactory preparations were available. The addition of enzymes to raw compostable materials is unnecessary because bacteria synthesize efficiently and rapidly all the enzymes required.

Another popular starter is "hormones," a term popularly used to designate the growth factors and vitamins needed by bacteria or other organisms. The organic constituents of mixed compostable materials usually contain all the growth factors needed for normal growth. Also, growth factors and vitamins are produced by micro-organisms and will undoubtedly be produced in sufficient quantities in a mixed microbial population to meet normal requirements.

A third group of additives are "biocatalyst" or "activated factors," various biological materials which are supposed to activate and accelerate decomposition and stabilization of organic material.

In some cases the "activator" supplies some material which is lacking in the compost. For example, straw or paper, which does not contain the necessary biological nutrients, is not readily composted alone, but if nitrogen and phosphorus are added, the straw and paper will serve as the carbon source for decomposition.

Agricultural experimentation has found starters or additives usually unnecessary. In one experiment, horse manure, compost material, normal soil, and special commercially prepared bacterial cultures were mixed with garbage and refuse to form one set of compost piles. In another set of piles, materials were composted without these special commercially prepared bacterial cultures. It was found that, although rich in bacteria, none of the inocula accelerated the composting process or improved the final product. There was no significant difference in the temperature curves or in the chemical analyses of the material at different intervals during the composting period. Therefore, it was concluded that when the environment is appropriate, the varied indigenous biological population will multiply rapidly and composting is not delayed.

COMPOSTING

Compost is organic matter that has been decomposed in a process called composting. This process recycles various organic materials otherwise regarded as waste products and produces a soil conditioner (the compost).

Compost is rich in nutrients. It is used, for example, in gardens, landscaping, horticulture, urban agriculture and organic farming. The compost itself is beneficial for the land in many ways, including as a soil conditioner, a fertilizer, addition of vital humus or humic acids, and as a natural pesticide for soil. In ecosystems, compost is useful for erosion control, land and stream reclamation, wetland construction, and as landfill cover.

At the simplest level, the process of composting requires making a heap of wet organic matter (also called green waste), such as leaves, grass, and food scraps, and waiting for the materials to break down into humus after a period of months. However, composting also can take place as a multi-step, closely monitored process with measured inputs of water, air, and carbon- and nitrogen-rich materials. The decomposition process is aided by shredding the plant matter, adding water and ensuring proper aeration by regularly turning the mixture when open piles or "windrows" are used. Earthworms and fungi further break up the material. Bacteria requiring oxygen to function (aerobic bacteria) and fungi manage the chemical process by converting the inputs into heat, carbon dioxide, and ammonium.

Fundamentals

Home compost barrel.

Composting is an aerobic method (meaning that it requires the presence of air) of decomposing organic solid wastes. It can therefore be used to recycle organic material. The process involves decomposition of organic material into a humus-like material, known as compost, which is a good fertilizer for plants. Composting requires the following three components: human management, aerobic conditions, development of internal biological heat.

Composting organisms require four equally important ingredients to work effectively:

- Carbon — For energy; the microbial oxidation of carbon produces the heat, if included at suggested levels. High carbon materials tend to be brown and dry.
- Nitrogen — To grow and reproduce more organisms to oxidize the carbon. High nitrogen materials tend to be green (or colorful, such as fruits and vegetables) and wet.

- Oxygen — For oxidizing the carbon, the decomposition process.
- Water — In the right amounts to maintain activity without causing anaerobic conditions.

Materials in a compost pile.

Certain ratios of these materials will provide microorganisms to work at a rate that will heat up the pile. Active management of the pile (e.g. turning) is needed to maintain sufficient supply of oxygen and the right moisture level. The air/water balance is critical to maintaining high temperatures (135°-160 °Fahrenheit/50° - 70 °Celsius) until the materials are broken down.

The most efficient composting occurs with an optimal carbon:nitrogen ratio of about 25:1. Hot container composting focuses on retaining the heat to increase decomposition rate and produce compost more quickly. Rapid composting is favored by having a C/N ratio of ~30 or less. Above 30 the substrate is nitrogen starved, below 15 it is likely to outgas a portion of nitrogen as ammonia.

Food scraps compost heap.

Nearly all plant and animal materials have both carbon and nitrogen, but amounts vary widely, with characteristics noted above (dry/wet, brown/green). Fresh grass clippings have an average ratio of about 15:1 and dry autumn leaves about 50:1 depending on species. Mixing equal parts by volume approximates the ideal C:N range. Few individual situations will provide the ideal mix of materials at any point. Observation of amounts, and consideration of different materials as a pile is built over time, can quickly achieve a workable technique for the individual situation.

Microorganisms

With the proper mixture of water, oxygen, carbon, and nitrogen, micro-organisms are able to break down organic matter to produce compost. The composting process is dependent on micro-organisms to break down organic matter into compost. There are many types of microorganisms found in active compost of which the most common are:

- Bacteria - The most numerous of all the microorganisms found in compost. Depending on the phase of composting, mesophilic or thermophilic bacteria may predominate.
- Actinobacteria - Necessary for breaking down paper products such as newspaper, bark, etc.
- Fungi - Molds and yeast help break down materials that bacteria cannot, especially lignin in woody material.
- Protozoa - Help consume bacteria, fungi and micro organic particulates.
- Rotifers - Rotifers help control populations of bacteria and small protozoans.

In addition, earthworms not only ingest partly composted material, but also continually re-create aeration and drainage tunnels as they move through the compost.

Phases of Composting

Three years old household compost.

Under ideal conditions, composting proceeds through three major phases:

- An initial, mesophilic phase, in which the decomposition is carried out under moderate temperatures by mesophilic microorganisms.
- As the temperature rises, a second, thermophilic phase starts, in which the decomposition is carried out by various thermophilic bacteria under high temperatures.
- As the supply of high-energy compounds dwindles, the temperature starts to decrease, and the mesophiles once again predominate in the maturation phase.

Slow and Rapid Composting

There are many proponents of rapid composting that attempt to correct some of the perceived problems associated with traditional, slow composting. Many advocate that compost can be made in 2 to 3 weeks. Many such short processes involve a few changes to traditional methods, including smaller, more homogenized pieces in the compost, controlling carbon-to-nitrogen ratio (C:N) at 30 to 1 or less, and monitoring the moisture level more carefully. However, none of these parameters differ significantly from the early writings of compost researchers, suggesting that in fact modern composting has not made significant advances over the traditional methods that take a few months to work. For this reason and others, many scientists who deal with carbon transformations are sceptical that there is a "super-charged" way to get nature to make compost rapidly.

Both sides may be right to some extent. The bacterial activity in rapid high heat methods breaks down the material to the extent that pathogens and seeds are destroyed, and the original feedstock is unrecognizable. At this stage, the compost can be used to prepare fields or other planting areas. However, most professionals recommend that the compost be given time to cure before using in a nursery for starting seeds or growing young plants. The curing time allows fungi to continue the decomposition process and eliminating phytotoxic substances.

An alternative approach is anaerobic fermentation, known as bokashi. It retains carbon bonds, is faster than decomposition, and for application to soil requires only rapid but thorough aeration rather than curing. It depends on sufficient carbohydrates in the treated material.

Pathogen Removal

Composting can destroy pathogens or unwanted seeds. Unwanted living plants (or weeds) can be discouraged by covering with mulch/compost. The "microbial pesticides" in compost may include thermophiles and mesophiles.

Thermophilic (high-temperature) composting is well known to destroy many seeds and nearly all types of pathogens (exceptions may include prions). The sanitizing qualities of (thermophilic) composting are desirable where there is a high likelihood of pathogens, such as with manure.

Materials that can be Composted

Composting is a process used for resource recovery. It can recycle an unwanted by-product from another process (a waste) into a useful new product.

Organic Solid Waste (Green Waste)

A large compost pile that is steaming with the heat generated by thermophilic microorganisms.

Composting is a process for converting decomposable organic materials into useful stable products. Therefore, valuable landfill space can be used for other wastes by composting these materials rather than dumping them on landfills. It may however be difficult to control inert and plastics contamination from municipal solid waste.

Co-composting is a technique that processes organic solid waste together with other input materials such as dewatered fecal sludge or sewage sludge.

Industrial composting systems are being installed to treat organic solid waste and recycle it rather than landfilling it. It is one example of an advanced waste processing system. Mechanical sorting of mixed waste streams combined with anaerobic digestion or in-vessel composting is called mechanical biological treatment. It is increasingly being used in developed countries due to regulations controlling the amount of organic matter allowed in landfills. Treating biodegradable waste before it enters a landfill reduces global warming from fugitive methane; untreated waste breaks down anaerobically in a landfill, producing landfill gas that contains methane, a potent greenhouse gas.

Animal Manure and Bedding

On many farms, the basic composting ingredients are animal manure generated on the farm and bedding. Straw and sawdust are common bedding materials. Non-traditional bedding materials are also used, including newspaper and chopped cardboard. The amount of manure composted on a livestock farm is often determined by cleaning schedules, land availability, and weather conditions. Each type of manure has its own physical, chemical, and biological characteristics. Cattle and horse manures, when mixed with bedding, possess good qualities for composting. Swine manure, which is very wet and usually not mixed with bedding material, must be mixed with straw or similar raw materials. Poultry manure also must be blended with carbonaceous materials - those low in nitrogen preferred, such as sawdust or straw.

Human Excreta and Sewage Sludge

Human excreta can also be added as an input to the composting process since human excreta is a nitrogen-rich organic material. It can be either composted directly, like in composting toilets, or indirectly (as sewage sludge), after it has undergone treatment in a sewage treatment plant.

Urine can be put on compost piles or directly used as fertilizer. Adding urine to compost can increase temperatures and therefore increase its ability to destroy pathogens and unwanted seeds. Unlike feces, urine does not attract disease-spreading flies (such as houseflies or blowflies), and it does not contain the most hardy of pathogens, such as parasitic worm eggs. Urine usually does not smell for long, particularly when it is fresh, diluted, or put on sorbents.

Uses

Compost can be used as an additive to soil, or other matrices such as coir and peat, as a tilth improver, supplying humus and nutrients. It provides a rich *growing medium*, or a porous, absorbent material that holds moisture and soluble minerals, providing the support and nutrients in which plants can flourish, although it is rarely used alone, being primarily mixed with soil, sand, grit, bark chips, vermiculite, perlite, or clay granules to produce loam. Compost can be tilled directly into the soil or growing medium to boost the level of organic matter and the overall fertility of the soil. Compost that is ready to be used as an additive is dark brown or even black with an earthy smell.

Generally, direct seeding into a compost is not recommended due to the speed with which it may dry and the possible presence of phytotoxins in immature compost that may inhibit germination, and the possible tie up of nitrogen by incompletely decomposed lignin. It is very common to see blends of 20–30% compost used for transplanting seedlings at cotyledon stage or later.

Compost can be used to increase plant immunity to diseases and pests.

Composting Technologies

Various approaches have been developed to handle different ingredients, locations, throughput and applications for the composted product.

Industrial-scale

Industrial-scale composting can be carried out in the form of in-vessel composting, aerated static pile composting, vermicomposting, or windrow composting.

Vermicomposting

Worms in a bin being harvested.

Vermicompost is the product or process of organic material degradation using various species of worms, usually red wigglers, white worms, and earthworms, to create a heterogeneous mixture of decomposing vegetable or food waste (excluding meat, dairy, fats, or oils), bedding materials, and vermicast. Vermicast, also known as worm castings, worm humus or worm manure, is the end-product of the breakdown of organic matter by species of earthworm.

Vermicomposting can also be applied for treatment of sewage sludge.

Composting Toilets

A composting toilet collects human excreta. These are added to a compost heap that can be located in a chamber below the toilet seat. Sawdust and straw or other carbon rich materials are usually added as well. Some composting toilets do not require water or electricity; others may. If they do not use water for flushing they fall into the category of dry toilets. Some composting toilet designs use urine diversion, others do not. When properly managed, they do not smell. The composting process in these toilets destroys pathogens to some extent. The amount of pathogen destruction depends on the temperature (mesophilic or thermophilic conditions) and composting time. Composting toilets with a large composting container (of the type Clivus Multrum and derivations of it) are popular in United States, Canada, Australia, New Zealand and Sweden. They are available as commercial products, as designs for self builders or as "design derivatives" which are marketed under various names.

Composting toilet with a seal in the lid in Germany.

Black Soldier Fly Larvae

Black soldier fly (*Hermetia illucens*) larvae are able to rapidly consume large amounts of organic material when kept at around 30 °C. Black soldier fly larvae can reduce the dry matter of the organic waste by 73% and convert 16-22% of the dry matter in the waste to biomass. The resulting compost still contains nutrients and can be used for biogas production, or further traditional composting or vermicomposting The larvae are rich in fat and protein, and can be used as, for example, animal feed or biodiesel production. Enthusiasts have experimented with a large number of different waste products.

Bokashi

Inside a recently started bokashi bin. Food scraps are raised on a perforated plate (to drain runoff) and are partly covered by a layer of bran inoculated with Lactobacilli.

Bokashi is not composting as defined earlier, rather an alternative technology. It ferments (rather than decomposes) the input organic matter and feeds the result to the soil food web (rather than producing a soil conditioner). The process involves adding *Lactobacilli* to the input in an airtight container kept at normal room temperature. These bacteria ferment carbohydrates to lactic acid, which preserves the input. After this is complete the preserve is mixed into soil, converting the lactic acid to pyruvate, which enables soil life to consume the result.

Bokashi is typically applied to food waste from households, workplaces and catering establishments, because such waste normally holds a good proportion of carbohydrates; it is also applied to other organic waste by supplementing carbohydrates. Household containers ("bokashi bins") typically give a batch size of 5-10 kilograms, accumulated over a few weeks. In horticultural settings batches can be orders of magnitude greater.

Bokashi offers several advantages:

- Fermentation retains all the original carbon and energy. (In comparison, composting loses at least 50% of these and 75% or more in amateur use; composting also loses nitrogen, a macronutrient of plants, by emitting ammonia and the potent greenhouse gas nitrous oxide).
- Virtually the full range of food waste is accepted, without the exclusions of composting. The exception is large bones.
- Being airtight, the container inherently traps smells, and when opened the smell of fermentation is far less offensive than decomposition. Hence bokashi bins usually operate indoors, in or near kitchens.
- Similarly the container neither attracts insect pests nor allows them ingress.
- The process is inherently hygienic because lactic acid is a natural bactericide and anti-pathogen; even its own fermentation is self-limiting.
- Both preservation and consumption complete within a few weeks rather than months.
- The preserve can be stored until needed, for example if ground is frozen or waterlogged.
- The increased activity of the soil food web improves the soil texture, especially by worm action - in effect this is in-soil vermicomposting.

The importance of the first advantage should not be underestimated: the mass of any ecosystem depends on the energy it captures. Plants depend upon the soil ecosystem making nutrients available within soil water. Therefore, the richer the ecosystem, the richer the plants. (Plants can also take up nutrients from added chemicals, but these are at odds with the purpose of composting).

Other Systems at Household Level

Hügelkultur (Raised Garden Beds or Mounds)

The practice of making raised garden beds or mounds filled with rotting wood is also called *hügelkultur* in German. It is in effect creating a nurse log that is covered with soil.

Benefits of hügelkultur garden beds include water retention and warming of soil. Buried wood acts like a sponge as it decomposes, able to capture water and store it for later use by crops planted on top of the *hügelkultur* bed.

An almost completed Hügelkultur bed; the bed does not have soil on it yet.

Compost Tea

Compost teas are defined as water extracts leached from composted materials. Compost teas are generally produced from adding one volume of compost to 4-10 volumes of water, but there has also been debate about the benefits of aerating the mixture. Field studies have shown the benefits of adding compost teas to crops due to the adding of organic matter, increased nutrient availability and increased microbial activity. They have also been shown to have an effect on plant pathogens.

Worm Hotels

Worm Hotel in Amsterdam.

Worm Hotels accommodate useful worm in ideal conditions.

Related Technologies

Organic ingredients intended for composting can also be used to generate biogas through anaerobic digestion. This process stabilizes organic material. The residual material, sometimes in combination with sewage sludge can be treated by a composting process before selling or giving away the compost.

Examples:

Large-scale composting systems are used by many urban areas around the world.

- The world's largest municipal co-composter for municipal solid waste (MSW) is the Edmonton Composting Facility in Edmonton, Alberta, Canada, which turns 220,000 tonnes of municipal solid waste and 22,500 dry tonnes of sewage sludge per year into 80,000 tonnes of compost. The facility is 38,690 m^2 (416,500 sq.ft.) in area, equivalent to 4½ Canadian football fields, and the operating structure is the largest stainless steel building in North America.
- In 2006, Qatar awarded Keppel Seghers Singapore, a subsidiary of Keppel Corporation, a contract to begin construction on a 275,000 tonne/year anaerobic digestion and composting plant licensed by Kompogas Switzerland. This plant, with 15 independent anaerobic digesters, will be the world's largest composting facility once fully operational in early 2011 and forms part of Qatar's Domestic Solid Waste Management Centre, the largest integrated waste management complex in the Middle East.

Edmonton Composting Facility.

- Another large municipal solid waste composter is the Lahore Composting Facility in Lahore, Pakistan, which has a capacity to convert 1,000 tonnes of municipal solid waste per day into compost. It also has a capacity to convert substantial portion of the intake into refuse-derived fuel (RDF) materials for further combustion use in several energy consuming industries across Pakistan, for example in cement manufacturing companies where it is used to heat cement kilns. This project has also been approved by the Executive Board of the United Nations Framework Convention on Climate Change for reducing methane emissions, and has been registered with a capacity of reducing 108,686 tonnes carbon dioxide equivalent per annum.
- Kew Gardens in London has one of the biggest non-commercial compost heaps in Europe.
- Compost is used as a soil amendment in organic farming.

Composting as a recognized practice dates to at least the early Roman Empire, and was mentioned as early as Cato the Elder's 160 BCE piece *De Agri Cultura*. Traditionally, composting involved piling organic materials until the next planting season, at which time the materials would have decayed enough to be ready for use in the soil. The advantage of this method is that little working time or effort is required from the composter and it fits in naturally with agricultural practices in temperate climates. Disadvantages (from the modern perspective) are that space is used for a whole year, some nutrients might be leached due to exposure to rainfall, and disease-producing organisms and insects may not be adequately controlled.

Compost Basket.

Composting was somewhat modernized beginning in the 1920s in Europe as a tool for organic farming. The first industrial station for the transformation of urban organic materials into compost was set up in Wels, Austria in the year 1921. Early frequent citations for propounding composting within farming are for the German-speaking world Rudolf Steiner, founder of a farming method called biodynamics, and Annie Francé-Harrar, who was appointed on behalf of the government in Mexico and supported the country 1950–1958 to set up a large humus organization in the fight against erosion and soil degradation.

In the English-speaking world it was Sir Albert Howard who worked extensively in India on sustainable practices and Lady Eve Balfour who was a huge proponent of composting. Composting was imported to America by various followers of these early European movements by the likes of J.I. Rodale (founder of Rodale Organic Gardening), E.E. Pfeiffer (who developed scientific practices in biodynamic farming), Paul Keene (founder of Walnut Acres in Pennsylvania), and Scott and Helen Nearing (who inspired the back-to-the-land movement of the 1960s). Coincidentally, some of the above met briefly in India - all were quite influential in the U.S. from the 1960s into the 1980s.

Vermicomposting

Vermicompost (vermi-compost, vermiculture) is the product of the composting process using various species of worms, usually red wigglers, white worms, and other earthworms, to create a mixture of decomposing vegetable or food waste, bedding materials, and vermicast.

Vermicast (also called worm castings, worm humus, worm manure, or worm feces) is the end-product of the breakdown of organic matter by earthworms. These castings have been shown to contain reduced levels of contaminants and a higher saturation of nutrients than the organic materials before vermicomposting. It contains water-soluble nutrients and is an excellent, nutrient-rich organic fertilizer and soil conditioner. It is used in farming and small scale sustainable, organic farming. It can also be applied for treatment of sewage. A variation of the process is vermifiltration (or vermidigestion) which is used to remove organic matter, pathogens and oxygen demand from wastewater or directly from blackwater of flush toilets.

Vermicomposting has gained popularity in both industrial and domestic settings because, as compared with conventional composting, it provides a way to treat organic wastes more quickly. In manure composting, it also generates products that have lower salinity levels.

Food waste with worms.

The earthworm species (or composting worms) most often used are red wigglers (*Eisenia fetida* or *Eisenia andrei*), though European nightcrawlers (*Eisenia hortensis* or *Dendrobaena veneta*) could also be used. Red wigglers are recommended by most vermicomposting experts, as they have some of the best appetites and breed very quickly. Users refer to European nightcrawlers by a variety of other names, including *dendrobaenas*, *dendras*, Dutch nightcrawlers, and Belgian nightcrawlers.

Containing water-soluble nutrients, vermicompost is a nutrient-rich organic fertilizer and soil conditioner in a form that is relatively easy for plants to absorb. Worm castings are sometimes used as an organic fertilizer. Because the earthworms grind and uniformly mix minerals in simple forms, plants need only minimal effort to obtain them. The worms' digestive systems create environments that allow certain species of microbes to thrive to help create a "living" soil environment for plants. The fraction of soil which has gone through the digestive tract of earthworms is called the drilosphere.

Design Considerations

Suitable Worm Species

One of the species most often used for composting is the red wiggler or tiger worm (*Eisenia fetida* or *Eisenia andrei*); *Lumbricus rubellus* (a.k.a. red earthworm or dilong (China)) is another breed of worm that can be used, but it does not adapt as well to the shallow compost bin as does *Eisenia fetida*. European nightcrawlers (*Eisenia hortensis*) may also be used. Users refer to European nightcrawlers by a variety of other names, including dendrobaenas, dendras, and nightcrawlers. African Nightcrawlers (*Eudrilus eugeniae*) are another set of popular composters. *Lumbricus terrestris* (a.k.a. Canadian nightcrawlers (US) or common earthworm (UK)) are not recommended, since they burrow deeper than most compost bins can accommodate.

Blueworms (Perionyx Excavatus)

These species commonly are found in organic-rich soils throughout Europe and North America and live in rotting vegetation, compost, and manure piles. They may be an invasive species in some areas. As they are shallow-dwelling and feed on decomposing plant matter in the soil, they adapt easily to living on food or plant waste in the confines of a worm bin.

Composting worms are available to order online, from nursery mail-order suppliers or angling shops where they are sold as bait. They can also be collected from compost and manure piles. These species are not the same worms that are found in ordinary soil or on pavement when the soil is flooded by water.

Large Scale

Large-scale vermicomposting is practiced in Canada, Italy, Japan, India, Malaysia, the Philippines, and the United States. The vermicompost may be used for farming, landscaping, to create compost tea, or for sale. Some of these operations produce worms for bait and/or home vermicomposting.

There are two main methods of large-scale vermiculture. Some systems use a windrow, which consists of bedding materials for the earthworms to live in and acts as a large bin; organic material is added to it. Although the windrow has no physical barriers to prevent worms from escaping, in theory they should not due to an abundance of organic matter for them to feed on. Often windrows are used on a concrete surface to prevent predators from gaining access to the worm population.

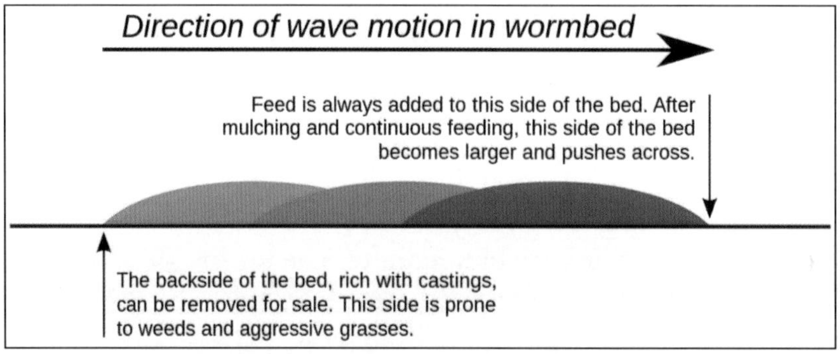

Movement of castings through a worm bed.

The windrow method and compost windrow turners were developed by Fletcher Sims Jr. of the Compost Corporation in Canyon, Texas. The Windrow Composting system is noted as a sustainable, cost-efficient way for farmers to manage dairy waste.

The second type of large-scale vermicomposting system is the raised bed or flow-through system. Here the worms are fed an inch of "worm chow" across the top of the bed, and an inch of castings are harvested from below by pulling a breaker bar across the large mesh screen which forms the base of the bed.

Because red worms are surface dwellers constantly moving towards the new food source, the flow-through system eliminates the need to separate worms from the castings before packaging. Flow-through systems are well suited to indoor facilities, making them the preferred choice for operations in colder climates.

Small Scale

Demonstration home scale worm bin at a community garden site (painted plywood).

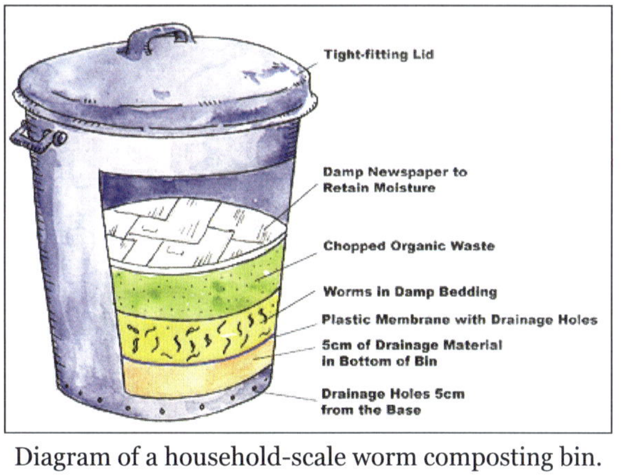

Diagram of a household-scale worm composting bin.

For vermicomposting at home, a large variety of bins are commercially available, or a variety of adapted containers may be used. They may be made of old plastic containers, wood, Styrofoam, or metal containers. The design of a small bin usually depends on where an individual wishes to store the bin and how they wish to feed the worms.

Some materials are less desirable than others in worm bin construction. Metal containers often conduct heat too readily, are prone to rusting, and may release heavy metals into the vermicompost.

Styrofoam containers may release chemicals into the organic material. Some cedars, Yellow cedar, and Redwood contain resinous oils that may harm worms, although Western Red Cedar has excellent longevity in composting conditions. Hemlock is another inexpensive and fairly rot-resistant wood species that may be used to build worm bins.

Bins need holes or mesh for aeration. Some people add a spout or holes in the bottom for excess liquid to drain into a tray for collection. The most common materials used are plastic: recycled polyethylene and polypropylene and wood. Worm compost bins made from plastic are ideal, but require more drainage than wooden ones because they are non-absorbent. However, wooden bins will eventually decay and need to be replaced.

Small-scale vermicomposting is well-suited to turn kitchen waste into high-quality soil amendments, where space is limited. Worms can decompose organic matter without the additional human physical effort (turning the bin) that bin composting requires.

Composting worms which are detritivorous (eaters of trash), such as the red wiggler *Eisenia fetidae*, are epigeic (surface dwellers) and together with symbiotic associated microbes are the ideal vectors for decomposing food waste. Common earthworms such as *Lumbricus terrestris* are anecic (deep burrowing) species and hence unsuitable for use in a closed system. Other soil species that contribute include insects, other worms and molds.

Climate and Temperature

There may be differences in vermicomposting method depending on the climate. It is necessary to monitor the temperatures of large-scale bin systems (which can have high heat-retentive properties), as the raw materials or feedstocks used can compost, heating up the worm bins as they decay and killing the worms.

The most common worms used in composting systems, redworms (*Eisenia foetida, Eisenia andrei,* and *Lumbricus rubellus*) feed most rapidly at temperatures of 15–25 °C (59-77 °F). They can survive at 10 °C (50 °F). Temperatures above 30 °C (86 °F) may harm them. This temperature range means that indoor vermicomposting with redworms is possible in all but tropical climates. Other worms like Perionyx excavatus are suitable for warmer climates. If a worm bin is kept outside, it should be placed in a sheltered position away from direct sunlight and insulated against frost in winter.

Feedstock

There are few food wastes that vermicomposting cannot compost, although meat waste and dairy products are likely to putrefy, and in outdoor bins can attract vermin. Green waste should be added in moderation to avoid heating the bin.

Small-scale or Home Systems

Such systems usually use kitchen and garden waste, using "earthworms and other microorganisms to digest organic wastes, such as kitchen scraps". This includes:

- All fruits and vegetables (including citrus, in limited quantities).

- Vegetable and fruit peels and ends.
- Coffee grounds and filters.
- Tea bags (even those with high tannin levels).
- Grains such as bread, cracker and cereal (including moldy and stale).
- Eggshells (rinsed off).
- Leaves and grass clippings (not sprayed with pesticides).
- Newspapers (most inks used in newspapers are not toxic).
- Paper toweling (which has not been used with cleaners or chemicals).

Large-scale or Commercial

Such vermicomposting systems need reliable sources of large quantities of food. Systems presently operating use:

- Dairy cow or pig manure Sewage sludge.
- Brewery waste.
- Cotton mill waste.
- Agricultural waste.
- Food processing and grocery waste.
- Cafeteria waste.
- Grass clippings and wood chips.

Harvesting

Worms in a bin being harvested.

Vermicompost is ready for harvest when it contains few-to-no scraps of uneaten food or bedding. There are several methods of harvesting from small-scale systems: "dump and hand sort", "let the worms do the sorting", "alternate containers" and "divide and dump." These differ on the amount of time and labor involved and whether the vermicomposter wants to save as many worms as possible from being trapped in the harvested compost.

The pyramid method of harvesting worm compost is commonly used in small-scale vermiculture, and is considered the simplest method for single layer bins. In this process, compost is separated into large clumps, which is placed back into composting for further breakdown, and lighter compost, with which the rest of the process continues. This lighter mix is placed into small piles on a tarp under the sunlight. The worms instinctively burrow to the bottom of the pile. After a few minutes, the top of the pyramid is removed repeatedly, until the worms are again visible. This repeats until the mound is composed mostly of worms.

When harvesting the compost, it is possible to separate eggs and cocoons and return them to the bin, thereby ensuring new worms are hatched. Cocoons are small, lemon-shaped yellowish objects that can usually be seen with the naked eye. The cocoons can hold up to 20 worms (though 2-3 is most common). Cocoons can lay dormant for as long as two years if conditions are not conducive for hatching.

Properties

Vermicompost has been shown to be richer in many nutrients than compost produced by other composting methods. It has also outperformed a commercial plant medium with nutrients added, but levels of magnesium required adjustment, as did pH.

However, in one study it has been found that homemade backyard vermicompost was lower in microbial biomass, soil microbial activity, and yield of a species of ryegrass than municipal compost.

It is rich in microbial life which converts nutrients already present in the soil into plant-available forms.

Unlike other compost, worm castings also contain worm mucus which helps prevent nutrients from washing away with the first watering and holds moisture better than plain soil.

Increases in the total nitrogen content in vermicompost, an increase in available nitrogen and phosphorus, as well as the increased removal of heavy metals from sludge and soil have been reported. The reduction in the bioavailability of heavy metals has been observed in a number of studies.

Benefits

Soil

- Improves soil aeration.
- Enriches soil with micro-organisms (adding enzymes such as phosphatase and cellulase).
- Microbial activity in worm castings is 10 to 20 times higher than in the soil and organic matter that the worm ingests.
- Attracts deep-burrowing earthworms already present in the soil.
- Improves water holding capacity.

Plant Growth

- Enhances germination, plant growth, and crop yield.
- Improves root growth and structure.
- Enriches soil with micro-organisms (adding plant hormones such as auxins and gibberellic acid).

Economic

- Biowastes conversion reduces waste flow to landfills.
- Elimination of biowastes from the waste stream reduces contamination of other recyclables collected in a single bin (a common problem in communities practicing Single-stream recycling).
- Creates low-skill jobs at local level.
- Low capital investment and relatively simple technologies make vermicomposting practical for less-developed agricultural regions.

Environmental

- Helps to close the "metabolic gap" through recycling waste on-site.
- Large systems often use temperature control and mechanized harvesting, however other equipment is relatively simple and does not wear out quickly.
- Production reduces greenhouse gas emissions such as methane and nitric oxide (produced in landfills or incinerators when not composted).

Uses

Soil Conditioner

Mid-scale worm bin (1 m X 2.5 m up to 1 m deep), freshly refilled with bedding.

Vermicompost can be mixed directly into the soil, or mixed with water to make a liquid fertilizer known as worm tea.

The dark brown waste liquid, or leachate, that drains into the bottom of some vermicomposting systems is not to be confused with worm tea.It is an uncomposted byproduct from when water-rich foods break down and may contain pathogens and toxins. It is best discarded or applied back to the bin when added moisture is needed for further processing.

The pH, nutrient, and microbial content of these fertilizers varies upon the inputs fed to worms. Pulverized limestone, or calcium carbonate can be added to the system to raise the pH.

Operation and Maintenance

Smells

Worms and fruit fly pupas under the lid of a home worm bin.

When closed, a well-maintained bin is odorless; when opened, it should have little smell—if any smell is present, it is earthy. The smell may also depend on the type of composted material added to the bin. An unhealthy worm bin may smell, potentially due to low oxygen conditions. Worms require gaseous oxygen. Oxygen can be provided by airholes in the bin, occasional stirring of bin contents, and removal of some bin contents if they become too deep or too wet. If decomposition becomes anaerobic from excess wet feedstock added to the bin, or the layers of food waste have become too deep, the bin will begin to smell of ammonia.

Moisture

Moisture must be maintained above 50%, as lower moisture content will not support worm respiration and can increase worm mortality. Operating moisture-content range should be between 70-90%, with a suggested content of 70-80% for vermicomposting-oriented vermiculture operations. If decomposition has become anaerobic, to restore healthy conditions and prevent the worms from dying, excess waste water must be reduced and the bin returned to a normal moisture level. To do this, first reduce addition of food scraps with a high moisture content and second, add fresh, dry bedding such as shredded newspaper to your bin, mixing it in well.

Pest Species

Pests such as rodents and flies are attracted by certain materials and odors, usually from large amounts of kitchen waste, particularly meat. Eliminating the use of meat or dairy product in a worm bin decreases the possibility of pests.

Predatory ants can be a problem in African countries.

In warm weather, fruit and vinegar flies breed in the bins if fruit and vegetable waste is not thoroughly covered with bedding. This problem can be avoided by thoroughly covering the waste by at least 5 centimetres (2.0 in) of bedding. Maintaining the correct pH (close to neutral) and water content of the bin (just enough water where squeezed bedding drips a couple of drops) can help avoid these pests as well.

Worms Escaping

Worms generally stay in the bin, but may try to leave the bin when first introduced, or often after a rainstorm when outside humidity is high. Maintaining adequate conditions in the worm bin and putting a light over the bin when first introducing worms should eliminate this problem.

Nutrient Levels

Commercial vermicomposters test, and may amend their products to produce consistent quality and results. Because the small-scale and home systems use a varied mix of feedstocks, the nitrogen, potassium and phosphorus content of the resulting vermicompost will also be inconsistent. NPK testing may be helpful before the vermicompost or tea is applied to the garden.

In order to avoid over-fertilization issues, such as nitrogen burn, vermicompost can be diluted as a tea 50:50 with water, or as a solid can be mixed in 50:50 with potting soil.

Additionally, the mucous layer created by worms which surrounds their castings allows for a "time release" effect, meaning not all nutrients are released at once. This also reduces the risk of burning the plants, as is common with the use and overuse of commercial fertilizers.

Application Examples

Vermicomposting (also known as vermiculture) is widely used in North America for on-site institutional processing of food scraps, such as in hospitals, universities, shopping malls, and correctional facilities. Vermicomposting is used for medium-scale on-site institutional organic material recycling, such as for food scraps from universities and shopping malls. It is selected either as a more environmentally friendly choice than conventional disposal, or to reduce the cost of commercial waste removal.

Researchers from the Pondicherry University discovered that worm composts can also be used to clean up heavy metals. The researchers found substantial reductions in heavy metals when the worms were released into the garbage and they are effective at removing lead, zinc, cadmium, copper and manganese.

AEROBIC DIGESTION

Aerobic digestion is a process in sewage treatment designed to reduce the volume of sewage sludge and make it suitable for subsequent use. More recently, technology has been developed that allows the treatment and reduction of other organic waste, such as food, cardboard and horticultural waste. It is a bacterial process occurring in the presence of oxygen. Bacteria rapidly consume organic matter and convert it into carbon dioxide, water and a range of lower molecular weight organic compounds. As there is no new supply of organic material from sewage, the activated sludge biota begin to die and are used as food by saprotrophic bacteria. This stage of the process is known as endogenous respiration and it is process that reduces the solid concentration in the sludge.

Process

Aerobic digestion is typically used in an activated sludge treatment plant. Waste activated sludge and primary sludge are combined, where appropriate, and passed to a thickener where the solids content is increased. This substantially reduces the volume that is required to be treated in the digester.The process is usually run as a batch process with more than one digester tank in operation at any one time. Air is pumped through the tank and the contents are stirred to keep the contents fully mixed. Carbon dioxide, waste air and small quantities of other gases including hydrogen sulfide are given off. These waste gases require treatment to reduce odours in works close to housing or capable of generating public nuisance. The digestion is continued until the percentage of degradable solids is reduced to between 20% and 10% depending on local conditions Where non sewage waste is being processed, organic waste such as food, cardboard and horticultural waste can be significantly reduced in volume leaving an output that can be used as soil improver or biomass fuel.

Advantages

Because the aerobic digestion occurs much faster than anaerobic digestion, the capital costs of aerobic digestion are lower.

The process is usually run at ambient temperature and the process is much less complex than anaerobic digestion and is easier to manage.

Disadvantages

The operating costs are typically much greater for aerobic digestion than for anaerobic digestion because of energy used by the blowers, pumps and motors needed to add oxygen to the process. However, recent technological advances include non-electrically aerated filter systems that use natural air currents for the aeration instead of electrically operated machinery.

The digested sludge is relatively low in residual energy and although it can be dried and incinerated to produce heat, the energy yield is very much lower than that produced by anaerobic digestion.

ATADs

Autothermal Thermophilic Aerobic Digestion (ATAD) is a faecal sludge treatment design concept that uses the nutrients in the sludge and the metabolic heat of the bacteria to create high temperatures in the aerobic digester. This gradually shifts the microbial community towards termophilic at temperatures typically at 55 degree Celsius or above. While the higher aeration requirements of ATADs further increase energy use and potential smell nuisance, the increased temperature makes the resulting biosolids much safer for re-use.

ANAEROBIC DIGESTION

Anaerobic digestion is a collection of processes by which microorganisms break down biodegradable material in the absence of oxygen. The process is used for industrial or domestic purposes to manage waste or to produce fuels. Much of the fermentation used industrially to produce food and drink products, as well as home fermentation, uses anaerobic digestion.

Anaerobic digestion occurs naturally in some soils and in lake and oceanic basin sediments, where it is usually referred to as "anaerobic activity". This is the source of marsh gas methane as discovered by Alessandro Volta in 1776.

The digestion process begins with bacterial hydrolysis of the input materials. Insoluble organic polymers, such as carbohydrates, are broken down to soluble derivatives that become available for other bacteria. Acidogenic bacteria then convert the sugars and amino acids into carbon dioxide, hydrogen, ammonia, and organic acids. These bacteria convert these resulting organic acids into acetic acid, along with additional ammonia, hydrogen, and carbon dioxide. Finally, methanogens convert these products to methane and carbon dioxide. The methanogenic archaea populations play an indispensable role in anaerobic wastewater treatments.

Anaerobic digestion is used as part of the process to treat biodegradable waste and sewage sludge. As part of an integrated waste management system, anaerobic digestion reduces the emission of landfill gas into the atmosphere. Anaerobic digesters can also be fed with purpose-grown energy crops, such as maize.

Anaerobic digestion is widely used as a source of renewable energy. The process produces a biogas, consisting of methane, carbon dioxide, and traces of other 'contaminant' gases. This biogas can be used directly as fuel, in combined heat and power gas engines or upgraded to natural gas-quality biomethane. The nutrient-rich digestate also produced can be used as fertilizer.

With the re-use of waste as a resource and new technological approaches that have lowered capital costs, anaerobic digestion has in recent years received increased attention among governments in a number of countries, among these the United Kingdom, Germany and Denmark.

Process

Many microorganisms affect anaerobic digestion, including acetic acid-forming bacteria (acetogens) and methane-forming archaea (methanogens). These organisms promote a number of chemical processes in converting the biomass to biogas.

Gaseous oxygen is excluded from the reactions by physical containment. Anaerobes utilize electron acceptors from sources other than oxygen gas. These acceptors can be the organic material itself or may be supplied by inorganic oxides from within the input material. When the oxygen source in an anaerobic system is derived from the organic material itself, the 'intermediate' end products are primarily alcohols, aldehydes, and organic acids, plus carbon dioxide. In the presence of specialised methanogens, the intermediates are converted to the 'final' end products of methane, carbon dioxide, and trace levels of hydrogen sulfide. In an anaerobic system, the majority of the chemical energy contained within the starting material is released by methanogenic bacteria as methane.

Populations of anaerobic microorganisms typically take a significant period of time to establish themselves to be fully effective. Therefore, common practice is to introduce anaerobic microorganisms from materials with existing populations, a process known as "seeding" the digesters, typically accomplished with the addition of sewage sludge or cattle slurry.

Process Stages

The four key stages of anaerobic digestion involve hydrolysis, acidogenesis, acetogenesis and methanogenesis. The overall process can be described by the chemical reaction, where organic material such as glucose is biochemically digested into carbon dioxide (CO_2) and methane (CH_4) by the anaerobic microorganisms.

$$C_6H_{12}O_6 \rightarrow 3CO_2 + 3CH_4$$

Hydrolysis

In most cases, biomass is made up of large organic polymers. For the bacteria in anaerobic digesters to access the energy potential of the material, these chains must first be broken down into their smaller constituent parts. These constituent parts, or monomers, such as sugars, are readily available to other bacteria. The process of breaking these chains and dissolving the smaller molecules into solution is called hydrolysis. Therefore, hydrolysis of these high-molecular-weight polymeric components is the necessary first step in anaerobic digestion. Through hydrolysis the complex organic molecules are broken down into simple sugars, amino acids, and fatty acids.

Acetate and hydrogen produced in the first stages can be used directly by methanogens. Other molecules, such as volatile fatty acids (VFAs) with a chain length greater than that of acetate must first be catabolised into compounds that can be directly used by methanogens.

Acidogenesis

The biological process of acidogenesis results in further breakdown of the remaining components by acidogenic (fermentative) bacteria. Here, VFAs are created, along with ammonia, carbon dioxide, and hydrogen sulfide, as well as other byproducts. The process of acidogenesis is similar to the way milk sours.

Acetogenesis

The third stage of anaerobic digestion is acetogenesis. Here, simple molecules created through the acidogenesis phase are further digested by acetogens to produce largely acetic acid, as well as carbon dioxide and hydrogen.

Methanogenesis

The terminal stage of anaerobic digestion is the biological process of methanogenesis. Here, methanogens use the intermediate products of the preceding stages and convert them into methane, carbon dioxide, and water. These components make up the majority of the biogas emitted from the system. Methanogenesis is sensitive to both high and low pHs and occurs between pH 6.5 and pH 8. The remaining, indigestible material the microbes cannot use and any dead bacterial remains constitute the digestate.

Configuration

Anaerobic digesters can be designed and engineered to operate using a number of different configurations and can be categorized into batch vs. continuous process mode, mesophilic vs. thermophilic temperature conditions, high vs. low portion of solids, and single stage vs. multistage processes. Continuous process requires more complex design, but still, it may be more economical than batch process, because batch process requires more initial building money and a larger volume of the digesters (spread across several batches) to handle the same amount of waste as a continuous process digester. Higher heat energy is required in a thermophilic system compared to a mesophilic system, but the thermophilic system requires much less time and has a larger gas output capacity and higher methane gas content, so one has to consider that trade-off carefully. For solids content, low will handle up to 15% solid content. Above this level is considered high solids content and can also be known as dry digestion. In a single stage process, one reactor houses the four anaerobic digestion steps. A multistage process utilizes two or more reactors for digestion to separate the methanogenesis and hydrolysis phases.

Batch or Continuous

Anaerobic digestion can be performed as a batch process or a continuous process. In a batch system, biomass is added to the reactor at the start of the process. The reactor is then sealed for the duration of the process. In its simplest form batch processing needs inoculation with already processed material to start the anaerobic digestion. In a typical scenario, biogas production will be formed with a normal distribution pattern over time. Operators can use this fact to determine when they believe the process of digestion of the organic matter has completed. There can be severe odour issues if a batch reactor is opened and emptied before the process is well completed. A more advanced type of batch approach has limited the odour issues by integrating anaerobic digestion with in-vessel composting. In this approach inoculation takes place through the use of recirculated degasified percolate. After anaerobic digestion has completed, the biomass is kept in the reactor which is then used for in-vessel composting before it is opened As the batch digestion is simple and requires less equipment and lower levels of design work, it is typically a cheaper form of digestion. Using more than one batch reactor at a plant can ensure constant production of biogas.

In continuous digestion processes, organic matter is constantly added (continuous complete mixed) or added in stages to the reactor (continuous plug flow; first in – first out). Here, the end products are constantly or periodically removed, resulting in constant production of biogas. A single or multiple digesters in sequence may be used. Examples of this form of anaerobic digestion include continuous stirred-tank reactors, upflow anaerobic sludge blankets, expanded granular sludge beds, and internal circulation reactors.

Temperature

The two conventional operational temperature levels for anaerobic digesters determine the species of methanogens in the digesters:

- Mesophilic digestion takes place optimally around 30 to 38 °C, or at ambient temperatures between 20 and 45 °C, where mesophiles are the primary microorganism present.

- Thermophilic digestion takes place optimally around 49 to 57 °C, or at elevated temperatures up to 70 °C, where thermophiles are the primary microorganisms present.

A limit case has been reached in Bolivia, with anaerobic digestion in temperature working conditions of less than 10 °C. The anaerobic process is very slow, taking more than three times the normal mesophilic time process. In experimental work at University of Alaska Fairbanks, a 1,000 litre digester using psychrophiles harvested from "mud from a frozen lake in Alaska" has produced 200–300 litres of methane per day, about 20 to 30% of the output from digesters in warmer climates. Mesophilic species outnumber thermophiles, and they are also more tolerant to changes in environmental conditions than thermophiles. Mesophilic systems are, therefore, considered to be more stable than thermophilic digestion systems. In contrast, while thermophilic digestion systems are considered less stable, their energy input is higher, with more biogas being removed from the organic matter in an equal amount of time. The increased temperatures facilitate faster reaction rates, and thus faster gas yields. Operation at higher temperatures facilitates greater pathogen reduction of the digestate. In countries where legislation, such as the Animal By-Products Regulations in the European Union, requires digestate to meet certain levels of pathogen reduction there may be a benefit to using thermophilic temperatures instead of mesophilic.

Additional pre-treatment can be used to reduce the necessary retention time to produce biogas. For example, certain processes shred the substrates to increase the surface area or use a thermal pretreatment stage (such as pasteurisation) to significantly enhance the biogas output. The pasteurisation process can also be used to reduce the pathogenic concentration in the digesate leaving the anaerobic digester. Pasteurisation may be achieved by heat treatment combined with maceration of the solids.

Solids Content

In a typical scenario, three different operational parameters are associated with the solids content of the feedstock to the digesters:

- High solids (dry—stackable substrate)
- High solids (wet—pumpable substrate)
- Low solids (wet—pumpable substrate)

High solids (dry) digesters are designed to process materials with a solids content between 25 and 40%. Unlike wet digesters that process pumpable slurries, high solids (dry – stackable substrate) digesters are designed to process solid substrates without the addition of water. The primary styles of dry digesters are continuous vertical plug flow and batch tunnel horizontal digesters. Continuous vertical plug flow digesters are upright, cylindrical tanks where feedstock is continuously fed into the top of the digester, and flows downward by gravity during digestion. In batch tunnel digesters, the feedstock is deposited in tunnel-like chambers with a gas-tight door. Neither approach has mixing inside the digester. The amount of pretreatment, such as contaminant removal, depends both upon the nature of the waste streams being processed and the desired quality of the digestate. Size reduction (grinding) is beneficial in continuous vertical systems, as it accelerates digestion, while batch systems avoid grinding and instead require structure (e.g. yard waste) to reduce compaction of the stacked pile. Continuous vertical dry digesters have a smaller footprint due to the shorter effective retention time and vertical design. Wet digesters can be designed to operate in either a high-solids content, with a total suspended solids (TSS) concentration greater than ~20%, or a low-solids concentration less than ~15%.

High solids (wet) digesters process a thick slurry that requires more energy input to move and process the feedstock. The thickness of the material may also lead to associated problems with abrasion. High solids digesters will typically have a lower land requirement due to the lower volumes associated with the moisture.High solids digesters also require correction of conventional performance calculations (e.g. gas production, retention time, kinetics, etc.) originally based on very dilute sewage digestion concepts, since larger fractions of the feedstock mass are potentially convertible to biogas.

Low solids (wet) digesters can transport material through the system using standard pumps that require significantly lower energy input. Low solids digesters require a larger amount of land than high solids due to the increased volumes associated with the increased liquid-to-feedstock ratio of the digesters. There are benefits associated with operation in a liquid environment, as it enables more thorough circulation of materials and contact between the bacteria and their food. This enables the bacteria to more readily access the substances on which they are feeding, and increases the rate of gas production.

Complexity

Digestion systems can be configured with different levels of complexity. In a single-stage digestion system (one-stage), all of the biological reactions occur within a single, sealed reactor or holding tank. Using a single stage reduces construction costs, but results in less control of the reactions occurring within the system. Acidogenic bacteria, through the production of acids, reduce the pH of the tank. Methanogenic bacteria, as outlined earlier, operate in a strictly defined pH range. Therefore, the biological reactions of the different species in a single-stage reactor can be in direct competition with each other. Another one-stage reaction system is an anaerobic lagoon. These lagoons are pond-like, earthen basins used for the treatment and long-term storage of manures. Here the anaerobic reactions are contained within the natural anaerobic sludge contained in the pool.

In a two-stage digestion system (multistage), different digestion vessels are optimised to bring maximum control over the bacterial communities living within the digesters. Acidogenic bacteria produce organic acids and more quickly grow and reproduce than methanogenic bacteria. Methanogenic bacteria require stable pH and temperature to optimise their performance.

Under typical circumstances, hydrolysis, acetogenesis, and acidogenesis occur within the first reaction vessel. The organic material is then heated to the required operational temperature (either mesophilic or thermophilic) prior to being pumped into a methanogenic reactor. The initial hydrolysis or acidogenesis tanks prior to the methanogenic reactor can provide a buffer to the rate at which feedstock is added. Some European countries require a degree of elevated heat treatment to kill harmful bacteria in the input waste. In this instance, there may be a pasteurisation or sterilisation stage prior to digestion or between the two digestion tanks. Notably, it is not possible to completely isolate the different reaction phases, and often some biogas is produced in the hydrolysis or acidogenesis tanks.

Residence Time

The residence time in a digester varies with the amount and type of feed material, and with the configuration of the digestion system. In a typical two-stage mesophilic digestion, residence time varies between 15 and 40 days, while for a single-stage thermophilic digestion, residence times is normally faster and takes around 14 days. The plug-flow nature of some of these systems will mean the full degradation of the material may not have been realised in this timescale. In this event, digestate exiting the system will be darker in colour and will typically have more odour.

In the case of an upflow anaerobic sludge blanket digestion (UASB), hydraulic residence times can be as short as 1 hour to 1 day, and solid retention times can be up to 90 days. In this manner, a UASB system is able to separate solids and hydraulic retention times with the use of a sludge blanket. Continuous digesters have mechanical or hydraulic devices, depending on the level of solids in the material, to mix the contents, enabling the bacteria and the food to be in contact. They also allow excess material to be continuously extracted to maintain a reasonably constant volume within the digestion tanks.

Inhibition

The anaerobic digestion process can be inhibited by several compounds, affecting one or more of the bacterial groups responsible for the different organic matter degradation steps. The degree of the inhibition depends, among other factors, on the concentration of the inhibitor in the digester. Potential inhibitors are ammonia, sulfide, light metal ions (Na, K, Mg, Ca, Al), heavy metals, some organics (chlorophenols, halogenated aliphatics, N-substituted aromatics, long chain fatty acids), etc.

In above figure, farm-based maize silage digester located near Neumünster in Germany, the green, inflatable biogas holder is shown on top of the digester. Two-stage, low solids, UASB digestion component of a mechanical biological treatment system near Tel Aviv; the process water is seen in balance tank and sequencing batch reactor.

Feedstocks

The most important initial issue when considering the application of anaerobic digestion systems is the feedstock to the process. Almost any organic material can be processed with anaerobic digestion; however, if biogas production is the aim, the level of putrescibility is the key factor in its successful application. The more putrescible (digestible) the material, the higher the gas yields possible from the system.

Anaerobic lagoon and generators at the Cal Poly Dairy.

Feedstocks can include biodegradable waste materials, such as waste paper, grass clippings, leftover food, sewage, and animal waste. Woody wastes are the exception, because they are largely unaffected by digestion, as most anaerobes are unable to degrade lignin. Xylophalgeous anaerobes (lignin consumers) or using high temperature pretreatment, such as pyrolysis, can be used to break lignin down. Anaerobic digesters can also be fed with specially grown energy crops, such as silage, for dedicated biogas production. In Germany and continental Europe, these facilities are referred to as "biogas" plants. A codigestion or cofermentation plant is typically an agricultural anaerobic digester that accepts two or more input materials for simultaneous digestion.

The length of time required for anaerobic digestion depends on the chemical complexity of the material. Material rich in easily digestible sugars breaks down quickly, whereas intact lignocellulosic material rich in cellulose and hemicellulose polymers can take much longer to break down. Anaerobic microorganisms are generally unable to break down lignin, the recalcitrant aromatic component of biomass.

Anaerobic digesters were originally designed for operation using sewage sludge and manures. Sewage and manure are not, however, the material with the most potential for anaerobic digestion, as the biodegradable material has already had much of the energy content taken out by the animals that produced it. Therefore, many digesters operate with codigestion of two or more types of feedstock. For example, in a farm-based digester that uses dairy manure as the primary feedstock, the gas production may be significantly increased by adding a second feedstock, e.g., grass and corn (typical on-farm feedstock), or various organic byproducts, such as slaughterhouse waste, fats, oils and grease from restaurants, organic household waste, etc. (typical off-site feedstock).

Digesters processing dedicated energy crops can achieve high levels of degradation and biogas production. Slurry-only systems are generally cheaper, but generate far less energy than those using crops, such as maize and grass silage; by using a modest amount of crop material (30%), an anaerobic digestion plant can increase energy output tenfold for only three times the capital cost, relative to a slurry-only system.

Moisture Content

A second consideration related to the feedstock is moisture content. Drier, stackable substrates, such as food and yard waste, are suitable for digestion in tunnel-like chambers. Tunnel-style systems typically have near-zero wastewater discharge, as well, so this style of system has advantages where the discharge of digester liquids are a liability. The wetter the material, the more suitable it will be to handling with standard pumps instead of energy-intensive concrete pumps and physical means of movement. Also, the wetter the material, the more volume and area it takes up relative to the levels of gas produced. The moisture content of the target feedstock will also affect what type of system is applied to its treatment. To use a high-solids anaerobic digester for dilute feedstocks, bulking agents, such as compost, should be applied to increase the solids content of the input material. Another key consideration is the carbon:nitrogen ratio of the input material. This ratio is the balance of food a microbe requires to grow; the optimal C:N ratio is 20–30:1. Excess N can lead to ammonia inhibition of digestion.

Contamination

The level of contamination of the feedstock material is a key consideration. If the feedstock to the digesters has significant levels of physical contaminants, such as plastic, glass, or metals, then processing to remove the contaminants will be required for the material to be used. If it is not removed, then the digesters can be blocked and will not function efficiently. It is with this understanding that mechanical biological treatment plants are designed. The higher the level of pre-treatment a feedstock requires, the more processing machinery will be required, and, hence, the project will have higher capital costs.

After sorting or screening to remove any physical contaminants from the feedstock, the material is often shredded, minced, and mechanically or hydraulically pulped to increase the surface area available to microbes in the digesters and, hence, increase the speed of digestion. The maceration of solids can be achieved by using a chopper pump to transfer the feedstock material into the air-tight digester, where anaerobic treatment takes place.

Substrate Composition

Substrate composition is a major factor in determining the methane yield and methane production rates from the digestion of biomass. Techniques to determine the compositional characteristics of the feedstock are available, while parameters such as solids, elemental, and organic analyses are important for digester design and operation. Methane yield can be estimated from the elemental composition of substrate along with an estimate of its degradability (the fraction of the substrate that is converted to biogas in a reactor). In order to predict biogas composition (the relative fractions of methane and carbon dioxide) it is necessary to estimate carbon dioxide partitioning between the aqueous and gas phases, which requires additional information (reactor temperature, pH, and substrate composition) and a chemical speciation model. Direct measurements of biomethanation potential are also made using gas evolution or more recent gravimetric assays.

Applications

Using anaerobic digestion technologies can help to reduce the emission of greenhouse gases in a number of key ways:

- Replacement of fossil fuels.
- Reducing or eliminating the energy footprint of waste treatment plants.

- Reducing methane emission from landfills.
- Displacing industrially produced chemical fertilizers.
- Reducing vehicle movements.
- Reducing electrical grid transportation losses.
- Reducing usage of LP Gas for cooking.
- An important component of the Zero Waste initiatives.

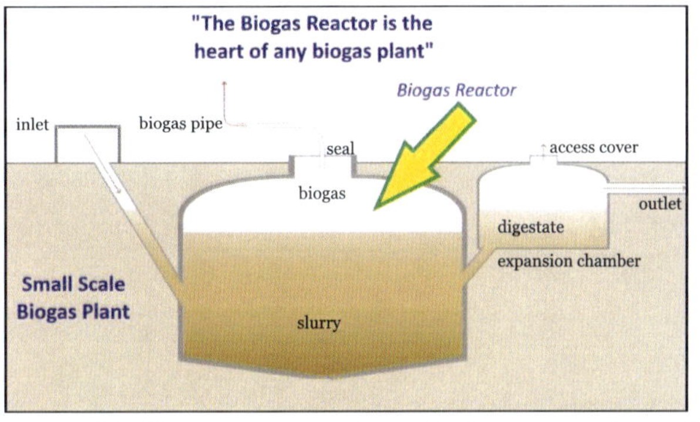

Schematic of an anaerobic digester as part of a sanitation system. It produces a digested slurry (digestate) that can be used as a fertilizer, and biogas that can be used for energy.

Waste and Wastewater Treatment

Anaerobic digesters in a sewage treatment plant. The methane gas is managed by burning through a gas flare.

Anaerobic digestion is particularly suited to organic material, and is commonly used for industrial effluent, wastewater and sewage sludge treatment. Anaerobic digestion, a simple process, can greatly reduce the amount of organic matter which might otherwise be destined to be dumped at sea, dumped in landfills, or burnt in incinerators.

Pressure from environmentally related legislation on solid waste disposal methods in developed countries has increased the application of anaerobic digestion as a process for reducing waste volumes and generating useful byproducts. It may either be used to process the source-separated fraction of municipal waste or alternatively combined with mechanical sorting systems, to process residual mixed municipal waste. These facilities are called mechanical biological treatment plants.

If the putrescible waste processed in anaerobic digesters were disposed of in a landfill, it would break down naturally and often anaerobically. In this case, the gas will eventually escape into the atmosphere. As methane is about 20 times more potent as a greenhouse gas than carbon dioxide, this has significant negative environmental effects.

In countries that collect household waste, the use of local anaerobic digestion facilities can help to reduce the amount of waste that requires transportation to centralized landfill sites or incineration facilities. This reduced burden on transportation reduces carbon emissions from the collection vehicles. If localized anaerobic digestion facilities are embedded within an electrical distribution network, they can help reduce the electrical losses associated with transporting electricity over a national grid.

Power Generation

In developing countries, simple home and farm-based anaerobic digestion systems offer the potential for low-cost energy for cooking and lighting. From 1975, China and India have both had large, government-backed schemes for adaptation of small biogas plants for use in the household for cooking and lighting. At present, projects for anaerobic digestion in the developing world can gain financial support through the United Nations Clean Development Mechanism if they are able to show they provide reduced carbon emissions.

Methane and power produced in anaerobic digestion facilities can be used to replace energy derived from fossil fuels, and hence reduce emissions of greenhouse gases, because the carbon in biodegradable material is part of a carbon cycle. The carbon released into the atmosphere from the combustion of biogas has been removed by plants for them to grow in the recent past, usually within the last decade, but more typically within the last growing season. If the plants are regrown, taking the carbon out of the atmosphere once more, the system will be carbon neutral. In contrast, carbon in fossil fuels has been sequestered in the earth for many millions of years, the combustion of which increases the overall levels of carbon dioxide in the atmosphere.

Biogas from sewage sludge treatment is sometimes used to run a gas engine to produce electrical power, some or all of which can be used to run the sewage works. Some waste heat from the engine is then used to heat the digester. The waste heat is, in general, enough to heat the digester to the required temperatures. The power potential from sewage works is limited – in the UK, there are about 80 MW total of such generation, with the potential to increase to 150 MW, which is insignificant compared to the average power demand in the UK of about 35,000 MW. The scope for biogas generation from nonsewage waste biological matter – energy crops, food waste, abattoir waste, etc. - is much higher, estimated to be capable of about 3,000 MW.Farm biogas plants using animal waste and energy crops are expected to contribute to reducing CO_2 emissions and strengthen the grid, while providing UK farmers with additional revenues.

Some countries offer incentives in the form of, for example, feed-in tariffs for feeding electricity onto the power grid to subsidize green energy production.

In Oakland, California at the East Bay Municipal Utility District's main wastewater treatment plant (EBMUD), food waste is currently codigested with primary and secondary municipal wastewater solids and other high-strength wastes. Compared to municipal wastewater solids digestion alone, food waste codigestion has many benefits. Anaerobic digestion of food waste pulp from the EBMUD food waste process provides a higher normalized energy benefit, compared to municipal wastewater solids: 730 to 1,300 kWh per dry ton of food waste applied compared to 560 to 940 kWh per dry ton of municipal wastewater solids applied.

Grid Injection

Biogas grid-injection is the injection of biogas into the natural gas grid. The raw biogas has to be previously upgraded to biomethane. This upgrading implies the removal of contaminants such as hydrogen sulphide or siloxanes, as well as the carbon dioxide. Several technologies are available for this purpose, the most widely implemented being pressure swing adsorption (PSA), water or amine scrubbing (absorption processes) and, in recent years, membrane separation. As an alternative, the electricity and the heat can be used for on-site generation, resulting in a reduction of losses in the transportation of energy. Typical energy losses in natural gas transmission systems range from 1–2%, whereas the current energy losses on a large electrical system range from 5–8%.

In October 2010, Didcot Sewage Works became the first in the UK to produce biomethane gas supplied to the national grid, for use in up to 200 homes in Oxfordshire. By 2017, UK electricity firm Ecotricity plan to have digester fed by locally sourced grass fueling 6000 homes.

Vehicle Fuel

After upgrading with the above-mentioned technologies, the biogas (transformed into biomethane) can be used as vehicle fuel in adapted vehicles. This use is very extensive in Sweden, where over 38,600 gas vehicles exist, and 60% of the vehicle gas is biomethane generated in anaerobic digestion plants.

Fertilizer and Soil Conditioner

The solid, fibrous component of the digested material can be used as a soil conditioner to increase the organic content of soils. Digester liquor can be used as a fertiliser to supply vital nutrients to soils instead of chemical fertilisers that require large amounts of energy to produce and transport. The use of manufactured fertilisers is, therefore, more carbon-intensive than the use of anaerobic digester liquor fertiliser. In countries such as Spain, where many soils are organically depleted, the markets for the digested solids can be equally as important as the biogas.

Cooking Gas

By using a bio-digester, which produces the bacteria required for decomposing, cooking gas is generated. The organic garbage like fallen leaves, kitchen waste, food waste etc. are fed into a crusher unit, where the mixture is conflated with a small amount of water. The mixture is then fed into the bio-digester, where the bacteria decomposes it to produce cooking gas. This gas is piped to kitchen stove. A 2 cubic meter bio-digester can produce 2 cubic meter of cooking gas. This is equivalent to 1 kg of LPG. The notable advantage of using a bio-digester is the sludge which is a rich organic manure.

Products

The three principal products of anaerobic digestion are biogas, digestate, and water.

Biogas

Typical composition of biogas		
Compound	Formula	%
Methane	CH_4	50–75
Carbon dioxide	CO_2	25–50
Nitrogen	N_2	0–10
Hydrogen	H_2	0–1
Hydrogen sulfide	H_2S	0–3
Oxygen	O_2	0–0

Biogas is the ultimate waste product of the bacteria feeding off the input biodegradable feedstock (the methanogenesis stage of anaerobic digestion is performed by archaea, a micro-organism on a distinctly different branch of the phylogenetic tree of life to bacteria), and is mostly methane and carbon dioxide, with a small amount hydrogen and trace hydrogen sulfide. (As-produced, biogas also contains water vapor, with the fractional water vapor volume a function of biogas temperature). Most of the biogas is produced during the middle of the digestion, after the bacterial population has grown, and tapers off as the putrescible material is exhausted. The gas is normally stored on top of the digester in an inflatable gas bubble or extracted and stored next to the facility in a gas holder.

The methane in biogas can be burned to produce both heat and electricity, usually with a reciprocating engine or microturbine often in a cogeneration arrangement where the electricity and waste heat generated are used to warm the digesters or to heat buildings. Excess electricity can be sold to suppliers or put into the local grid. Electricity produced by anaerobic digesters is considered to be renewable energy and may attract subsidies. Biogas does not contribute to increasing atmospheric carbon dioxide concentrations because the gas is not released directly into the atmosphere and the carbon dioxide comes from an organic source with a short carbon cycle.

Biogas holder with lightning protection rods and backup gas flare.

Biogas may require treatment or 'scrubbing' to refine it for use as a fuel. Hydrogen sulfide, a toxic product formed from sulfates in the feedstock, is released as a trace component of the biogas. National environmental enforcement agencies, such as the U.S. Environmental Protection Agency or the English and Welsh Environment Agency, put strict limits on the levels of gases containing hydrogen sulfide, and, if the levels of hydrogen sulfide in the gas are high, gas scrubbing and cleaning equipment (such as amine gas treating) will be needed to process the biogas to within regionally accepted levels. Alternatively, the addition of ferrous chloride $FeCl_2$ to the digestion tanks inhibits hydrogen sulfide production.

Volatile siloxanes can also contaminate the biogas; such compounds are frequently found in household waste and wastewater. In digestion facilities accepting these materials as a component of the feedstock, low-molecular-weight siloxanes volatilise into biogas. When this gas is combusted in a gas engine, turbine, or boiler, siloxanes are converted into silicon dioxide (SiO_2), which deposits internally in the machine, increasing wear and tear. Practical and cost-effective technologies to remove siloxanes and other biogas contaminants are available at the present time. In certain applications, *in situ* treatment can be used to increase the methane purity by reducing the offgas carbon dioxide content, purging the majority of it in a secondary reactor.

In countries such as Switzerland, Germany, and Sweden, the methane in the biogas may be compressed for it to be used as a vehicle transportation fuel or input directly into the gas mains. In countries where the driver for the use of anaerobic digestion are renewable electricity subsidies, this route of treatment is less likely, as energy is required in this processing stage and reduces the overall levels available to sell.

Biogas carrying pipes.

Digestate

Digestate is the solid remnants of the original input material to the digesters that the microbes cannot use. It also consists of the mineralised remains of the dead bacteria from within the digesters. Digestate can come in three forms: fibrous, liquor, or a sludge-based combination of the two fractions. In two-stage systems, different forms of digestate come from different digestion tanks. In single-stage digestion systems, the two fractions will be combined and, if desired, separated by further processing.

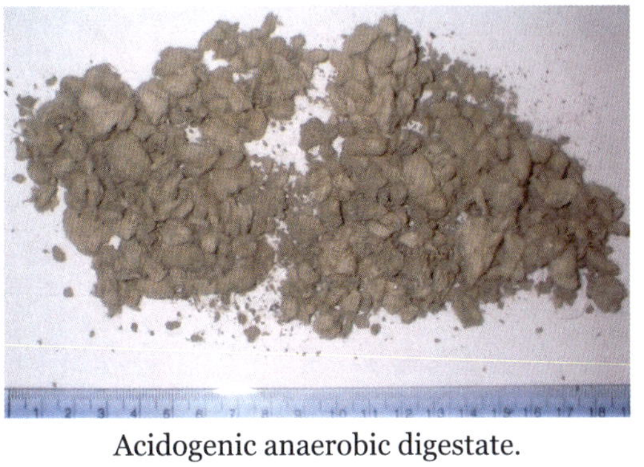
Acidogenic anaerobic digestate.

The second byproduct (acidogenic digestate) is a stable, organic material consisting largely of lignin and cellulose, but also of a variety of mineral components in a matrix of dead bacterial cells; some plastic may be present. The material resembles domestic compost and can be used as such or to make low-grade building products, such as fibreboard. The solid digestate can also be used as feedstock for ethanol production.

The third byproduct is a liquid (methanogenic digestate) rich in nutrients, which can be used as a fertiliser, depending on the quality of the material being digested. Levels of potentially toxic elements (PTEs) should be chemically assessed. This will depend upon the quality of the original feedstock. In the case of most clean and source-separated biodegradable waste streams, the levels of PTEs will be low. In the case of wastes originating from industry, the levels of PTEs may be higher and will need to be taken into consideration when determining a suitable end use for the material.

Digestate typically contains elements, such as lignin, that cannot be broken down by the anaerobic microorganisms. Also, the digestate may contain ammonia that is phytotoxic, and may hamper the growth of plants if it is used as a soil-improving material. For these two reasons, a maturation or composting stage may be employed after digestion. Lignin and other materials are available for degradation by aerobic microorganisms, such as fungi, helping reduce the overall volume of the material for transport. During this maturation, the ammonia will be oxidized into nitrates, improving the fertility of the material and making it more suitable as a soil improver. Large composting stages are typically used by dry anaerobic digestion technologies.

Wastewater

The final output from anaerobic digestion systems is water, which originates both from the moisture content of the original waste that was treated and water produced during the microbial reactions in the digestion systems. This water may be released from the dewatering of the digestate or may be implicitly separate from the digestate.

Gas street lamp.

The wastewater exiting the anaerobic digestion facility will typically have elevated levels of biochemical oxygen demand (BOD) and chemical oxygen demand (COD). These measures of the reactivity of the effluent indicate an ability to pollute. Some of this material is termed 'hard COD', meaning it cannot be accessed by the anaerobic bacteria for conversion into biogas. If this effluent were put directly into watercourses, it would negatively affect them by causing eutrophication. As such, further treatment of the wastewater is often required. This treatment will typically be an oxidation stage wherein air is passed through the water in a sequencing batch reactors or reverse osmosis unit.

The history of anaerobic digestion is a long one, beginning as early as tenth century BCE in Assyria where biogas was used to heat bath water. Reported scientific interest in the manufacturing of gas produced by the natural decomposition of organic matter dates from the 17th century, when Robert Boyle and Stephen Hales noted that disturbing the sediment of streams and lakes released flammable gas. In 1778, the Italian physicist Alessandro Volta, the father of Electrochemistry, scientifically identified that gas as methane.

In 1808 Sir Humphry Davy proved the presence of methane in the gases produced by cattle manure. The first known anaerobic digester was built in 1859 at a leper colony in Bombay in India. In 1895, the technology was developed in Exeter, England, where a septic tank was used to generate gas for the sewer gas destructor lamp, a type of gas lighting. Also in England, in 1904, the first dual-purpose tank for both sedimentation and sludge treatment was installed in Hampton, London.

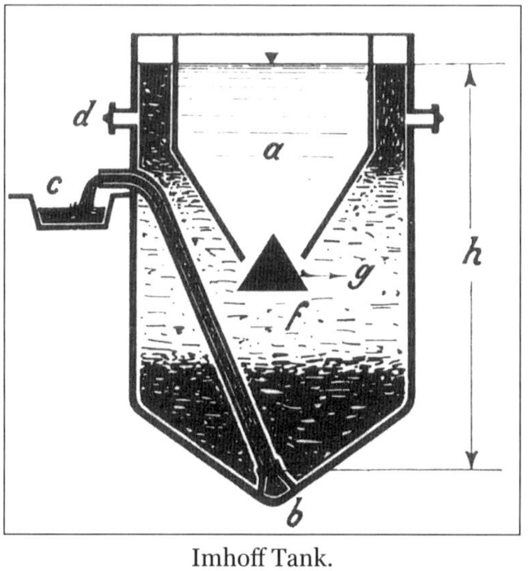

Imhoff Tank.

By the early 20th century, anaerobic digestion systems began to resemble the technology as it appears today. In 1906, Karl Imhoff created the Imhoff tank; an early form of anaerobic digester and model wastewater treatment system throughout the early 20th century. After 1920, closed tank systems began to replace the previously common use of anaerobic lagoons- covered earthen basins used to treat volatile solids. Research on anaerobic digestion began in earnest in the 1930s.

Around the time of World War I, production from biofuels slowed as petroleum production increased and its uses were identified. While fuel shortages during World War II re-popularized anaerobic digestion, interest in the technology decreased again after the war ended. Similarly, the 1970s energy crisis sparked interest in anaerobic digestion. In addition to high energy prices, factors affecting the adoption of Anaerobic Digestion systems include receptivity to innovation, pollution penalties, policy incentives, and the availability of subsidies and funding opportunities.

Today, anaerobic digesters are commonly found alongside farms to reduce nitrogen run-off from manure, or wastewater treatment facilities to reduce the costs of sludge disposal. Agricultural anaerobic digestion for energy production has become most popular in Germany, where there were 8,625 digesters in 2014. In the United Kingdom, there were 259 facilities by 2014, and 500 projects planned to become operational by 2019. In the United States, there were 191 operational plants across 34 states in 2012. Policy may explain why adoption rates are so different across these countries.

Feed-in tariffs in Germany were enacted in 1991, also known as FIT, providing long-term contracts compensating investments in renewable energy generation. Consequently, between 1991 and 1998 the number of anaerobic digester plants in Germany grew from 20 to 517. In the late 1990s, energy prices in Germany varied and investors became unsure of the market's potential. The German government responded by amending FIT four times between 2000 and 2011, increasing tariffs and improving the profitability of anaerobic digestion, and resulting in reliable returns for biogas production and continued high adoption rates across the country.

Hydrolysis

This is the first step in anaerobic digestion process, it involves the enzyme-mediated transformation of insoluble organic materials and higher molecular mass compounds such as lipids, polyssacharides, proteins, fats, nucleic acid etc. into soluble organic materials i.e. to compounds suitable for the use as source of energy and cell carbon such as monosaccharides, amino acids and other simple organic compounds. This step is carried out by strict anaerobes such as bacterides, clostridia and facultative bacteria such as streptococci etc. This first stage is very important because large organic molecules are simply too large to be directly absorbed and used by microorganisms as a substrate/food source. To accomplish biodegradation, certain microorganisms secrete different types of enzymes, called extracellular enzymes, which "cut" the larger molecules up into smaller pieces that the microorganism can then take into the cell and use as a source of energy and nutrition. Some microorganisms secrete several different enzymes, which allow them to break down different types of organic materials. Other microorganisms are specialised. For example, they secrete enzymes that break down either sugar or protein. Microorganisms that break down different sugars are called saccharolytic, while those that break down proteins are called proteolytic. There are different enzymes for sugars, proteins, fats etc. Table contains examples of some different groups of extracellular enzymes. Each group contains several enzymes that are specialised in various substrates, such as different proteins. The rate of decomposition during the hydrolysis stage depends greatly on the nature of the substrate. The transformation of cellulose and hemicellulose generally takes place more slowly than the decomposition of proteins.

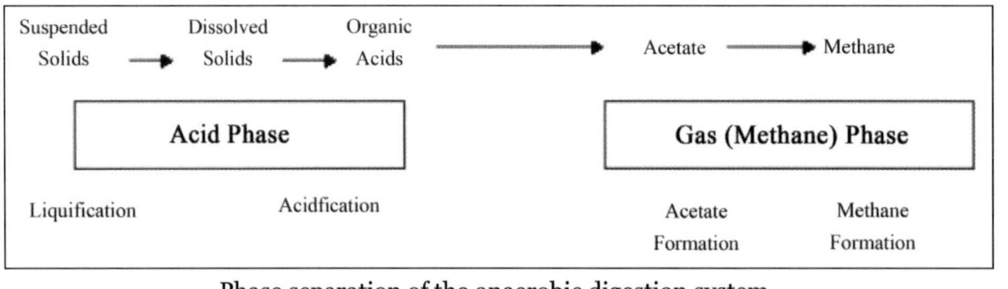

Phase separation of the anaerobic digestion system.

Table: Some important groups of hydrolytic enzymes and their functions.

Enzymes	Substrates	Breakdown products
Proteinase	Proteins	Amino acids
Cellulase	Cellulose	Cellobiose and glucose
Hemicellulase	Hemicellulose	Sugars, such as glucose, xylose, mannose and arabinose
Amylase	Starch	Glucose
Lipase	Fats	Fatty acids and glycerol
Pectinase	Pectin	Sugars, such as galactose, arabinose and polygalacticuronic acid

Acidogenesis

Acidogenesis is the second stage in the four stages of anaerobic digestion:

- Hydrolysis: A chemical reaction where particulates are solubilized and large polymers converted into simpler monomers;
- Acidogenesis: A biological reaction where simple monomers are converted into volatile fatty acids;
- Acetogenesis: A biological reaction where volatile fatty acids are converted into acetic acid, carbon dioxide, and hydrogen;
- Methanogenesis: A biological reaction where acetates are converted into methane and carbon dioxide, while hydrogen is consumed.

Anaerobic digestion is a complex biochemical process of biologically mediated reactions by a consortium of microorganisms to convert organic compounds into methane and carbon dioxide. It is a stabilization process, reducing odor, pathogens, and mass reduction.

Hydrolytic bacteria form a variety of reduced end-products from the fermentation of a given substrate. One fundamental question that arises concerns the metabolic features that control carbon and electron flow to a given reduced end-product during pure culture and mixed methanogenic cultures of hydrolytic bacteria. *Thermoanaerobium brockii* is a representative thermophilic, hydrolytic bacterium, which ferments glucose, via the Embden–Meyerhof Parnas Pathway. *T. brockii* is an atypical hetero-lactic acid bacterium because it forms molecular hydrogen (H_2), in addition to lactic acid and ethanol. The reduced end-products of glucose fermentation are enzymatically formed from pyruvate, via the following mechanisms: lactate by [[Fructose bisphosphatase|fructose 1-6]] all-phosphate (F6P) activated lactate dehydrogenase; H2 by pyruvate ferredoxin oxidoreductase and hydrogenase; and ethanol via NADH- and NADPH-linked alcohol dehydrogenase.

By its side, the acidogenic activity was found in the early 20th century, but it was not until the mid-1960s that the engineering of phases separation was assumed in order to improve the stability and waste digesters treatment. In this phase, complex molecules (carbohydrates, lipids, and proteins) are depolymerized into soluble compounds by hydrolytic enzymes (cellulases, hemicellulases, amylases, lipases and proteases). The hydrolyzed compounds are fermented into volatile fatty acids (acetate, propionate, butyrate, and lactate), neutral compounds (ethanol, methanol), ammonia, hydrogen and carbon dioxide.

Acetogenesis is one of the main reactions of this stage, in this, the intermediary metabolites produced are metabolized to acetate, hydrogen and carbonic gas by the three main groups of bacteria:

- Homoacetogens;
- Syntrophes;
- Sulphoreductors.

For the acetic acid production are considered three kind of bacteria:

- Clostridium aceticum;
- Acetobacter woodii;
- Clostridium termoautotrophicum.

Winter y Wolfe, in 1979, demonstrated that *A. wodii* in syntrophic association with Methanosarcina produce methane and carbon dioxide from fructose, instead of three molecules of acetate. Moorella thermoacetica and Clostridium formiaceticum are able to reduce the carbonic gas to acetate, but they do not have hydrogenases which inhabilite the hydrogen use, so they can produce three molecules of acetate from fructose. Acetic acid is equally a co-metabolite of the organic substrates fermentation (sugars, glycerol, lactic acid, etc.) by diverse groups of microorganisms which produce different acids:

- Propionic bacteria (propionate + acetate);
- Clostridium (butyrate + acetate);
- Enterobacteria (acetate + lactate);
- Hetero-fermentative bacteria (acetate, propionate, butyrate, valerate, etc.).

Acetogenesis

The products produced in the acidogenic phase are consumed as substrates for the other microorganisms, active in the third phase. In the third phase, also called the acidogenic phase anaerobic oxidationare performed. Products which cannot be directly converted to methane by methanogenic bacteria are converted into methanogenic substrates, volatile fatty acids and alcohols (VFA) are oxidized into methanogenic substrates like acetate, hydrogen and carbon dioxide, VFA with carbon chains longer than one unit are oxidized into acetate and hydrogen. It is important that the organisms which carry out the anaerobic oxidation reactions collaborate with the next group, the methane forming microorganisms; this collaboration depends on the partial pressure of the hydrogen present in the system. Under anaerobic oxidation, protons are used as the final electron acceptors which lead to the production of H2. However these oxidation reactions can only occur if the partial pressure of hydrogen is low, which explains why the collaboration with the methanogens is very important since they will continuously consume the H2, to produce methane. Hence during this symbiotic relationship inter-species hydrogen transfer occurs.

Methanogenesis

Methanogenesis or biomethanation is the formation of methane by microbes known as methanogens. Organisms capable of producing methane have been identified only from the domain Archaea, a group phylogenetically distinct from both eukaryotes and bacteria, although many live in close association with anaerobic bacteria. The production of methane is an important and widespread form of microbial metabolism. In anoxic environments, it is the final step in the decomposition of biomass. Methanogenesis is responsible for significant amounts of natural gas accumulations, the remainder being thermogenic.

Biochemistry

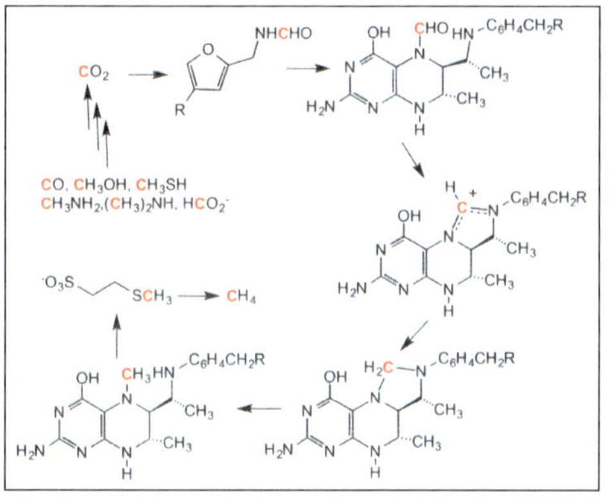

Cycle for methanogenesis, showing intermediates.

Methanogenesis in microbes is a form of anaerobic respiration. Methanogens do not use oxygen to respire; in fact, oxygen inhibits the growth of methanogens. The terminal electron acceptor in methanogenesis is not oxygen, but carbon. The carbon can occur in a small number of organic compounds, all with low molecular weights. The two best described pathways involve the use of acetic acid or inorganic carbon dioxide as terminal electron acceptors:

$$CO_2 + 4\,H_2 \rightarrow CH_4 + 2\,H_2O$$

$$CH_3COOH \rightarrow CH_4 + CO_2$$

During anaerobic respiration of carbohydrates, H_2 and acetate are formed in a ratio of 2:1 or lower, so H_2 contributes only ca. 33% to methanogenesis, with acetate contributing the greater proportion. In some circumstances, for instance in the rumen, where acetate is largely absorbed into the bloodstream of the host, the contribution of H_2 to methanogenesis is greater.

However, depending on pH and temperature, methanogenesis has been shown to use carbon from other small organic compounds, such as formic acid (formate), methanol, methylamines, tetramethylammonium, dimethyl sulfide, and methanethiol. The catabolism of the methyl compounds is mediated by methyl transferases to give methyl coenzyme M.

Proposed Mechanism

The biochemistry of methanogenesis involves the following coenzymes and cofactors: F420, coenzyme B, coenzyme M, methanofuran, and methanopterin.

The mechanism for the conversion of CH_3–S bond into methane involves a ternary complex of methyl coenzyme M and coenzyme B fit into a channel terminated by the axial site on nickel of the cofactor F430. One proposed mechanism invokes electron transfer from Ni(I) (to give Ni(II)), which initiates formation of CH_4. Coupling of the coenzyme M thiyl radical (RS·) with HS coenzyme B releases a proton and re-reduces Ni(II) by one-electron, regenerating Ni(I).

Reverse Methanogenesis

Some organisms can oxidize methane, functionally reversing the process of methanogenesis, also referred to as the anaerobic oxidation of methane (AOM). Organisms performing AOM have been found in multiple marine and freshwater environments including methane seeps, hydrothermal vents, coastal sediments and sulfate-methane transition zones. These organisms may accomplish reverse methanogenesis using a nickel-containing protein similar to methyl-coenzyme M reductase used by methanogenic archaea. Reverse methanogenesis occurs according to the reaction:

$$SO_4^{2-} + CH_4 \rightarrow HCO_3^- + HS^- + H_2O$$

Importance in Carbon Cycle

Methanogenesis is the final step in the decay of organic matter. During the decay process, electron acceptors (such as oxygen, ferric iron, sulfate, and nitrate) become depleted, while hydrogen (H_2) and carbon dioxide accumulate. Light organics produced by fermentation also accumulate. During advanced stages of organic decay, all electron acceptors become depleted except carbon dioxide. Carbon dioxide is a product of most catabolic processes, so it is not depleted like other potential electron acceptors.

Only methanogenesis and fermentation can occur in the absence of electron acceptors other than carbon. Fermentation only allows the breakdown of larger organic compounds, and produces small organic compounds. Methanogenesis effectively removes the semi-final products of decay: hydrogen, small organics, and carbon dioxide. Without methanogenesis, a great deal of carbon (in the form of fermentation products) would accumulate in anaerobic environments.

Natural Occurrence

In Ruminants

Testing Australian sheep for exhaled methane production.

Enteric fermentation occurs in the gut of some animals, especially ruminants. In the rumen, anaerobic organisms, including methanogens, digest cellulose into forms nutritious to the animal. Without these microorganisms, animals such as cattle would not be able to consume grasses.

The useful products of methanogenesis are absorbed by the gut, but methane is released from the animal mainly by belching (eructation). The average cow emits around 250 liters of methane per day. In this way, ruminants contribute about 25% of anthropogenic methane emissions. One method of methane production control in ruminants is by feeding them 3-nitrooxypropanol.

In Humans

Some humans produce flatus that contains methane. In one study of the feces of nine adults, five of the samples contained archaea capable of producing methane. Similar results are found in samples of gas obtained from within the rectum.

Even among humans whose flatus does contain methane, the amount is in the range of 10% or less of the total amount of gas.

In Plants

Many experiments have suggested that leaf tissues of living plants emit methane. Other research has indicated that the plants are not actually generating methane; they are just absorbing methane from the soil and then emitting it through their leaf tissues.

In Soils

Methanogens are observed in anoxic soil environments, contributing to the degradation of organic matter. This organic matter may be placed by humans through landfill, buried as sediment on the bottom of lakes or oceans as sediments, and as residual organic matter from sediments that have formed into sedimentary rocks.

Atmospheric methane is an important greenhouse gas with a global warming potential 25 times greater than carbon dioxide (averaged over 100 years), and methanogenesis in livestock and the decay of organic material is thus a considerable contributor to global warming. It may not be a net contributor in the sense that it works on organic material which used up atmospheric carbon dioxide when it was created, but its overall effect is to convert the carbon dioxide into methane which is a much more potent greenhouse gas.

Methanogenesis can also be beneficially exploited, to treat organic waste, to produce useful compounds, and the methane can be collected and used as biogas, a fuel. It is the primary pathway whereby most organic matter disposed of via landfill is broken down.

Extra-terrestrial Life

The presence of atmospheric methane has a role in the scientific search for extra-terrestrial life. The justification is that methane in the atmosphere will eventually dissipate, unless something is replenishing it. If methane is detected (by using a spectrometer for example) this may indicate that life is, or recently was, present. This was debated when methane was discovered in the Martian atmosphere by M.J. Mumma of NASA's Goddard Flight Center, and verified by the Mars Express Orbiter and in Titan's atmosphere by the Huygens probe. This debate was furthered with the discovery of 'transient', 'spikes of methane' on Mars by the Curiosity Rover.

It is also argued that atmospheric methane can come from volcanoes or other fissures in the planet's crust and that without an isotopic signature, the origin or source may be difficult to identify.

On 13 April 2017, NASA confirmed that the dive of the Cassini orbiter spacecraft on 28 October 2015 discovered the Enceladus plume which has all the ingredients for methanogenesis-based life forms to feed from. Previous results, published in March 2015, suggested hot water is interacting with rock beneath the sea; the new findings support that conclusion and add that the rock appears to be reacting chemically. From these observations scientists have determined that nearly 98 percent of the gas in the plume is water, about 1 percent is hydrogen and the rest is a mixture of other molecules including carbon dioxide, methane and ammonia.

Permafrost degradation makes the largest future transfer of methane.

Digestate

Digestate is the material remaining after the anaerobic digestion of a biodegradable feedstock. Anaerobic digestion produces two main products: digestate and biogas. Digestate is produced both by acidogenesis and methanogenesis and each has different characteristics.

Acidogenic Digestate

Acidogenic digestate is fibrous and consists of structural plant matter including lignin and cellulose. Acidogenic digestate has high moisture retention properties. The digestate may also contain minerals and remnants of bacteria.

Methanogenic Digestate

Methanogenic digestate is a sludge (sometimes called a liquor). This is often high in nutrients such as ammoniums and phosphates.

Uses

The primary use of digestate is as a soil conditioner. Acidogenic digestate provides moisture retention and organic content for soils. This organic material can break down further, aerobically in soil. Methanogenic digestate provides nutrients for plant growth. It can also be used to protect soils against erosion.

Acidogenic digestate can also be used as an environmentally friendly filler to give structure to composite plastics.

Growth trials on digestate originating from mixed waste have showed healthy growth results for crops. Digestate can also be used in intensive greenhouse cultivation of plants, e.g. in Digeponics.

Application of digestate has been shown to inhibit plant diseases and induction of resistance. Digestate application has a direct effect on soil-born diseases, and an indirect effect by stimulation of biological activity.

Digestate and Compost

Digestate is technically not compost although it is similar to it in physical and chemical characteristics. Compost is produced by aerobic digestion- decomposition by aerobes. This includes fungi and bacteria which are able to break down the lignin and cellulose to a greater extent.

Treatment, e.g. by ultrasonication, has shown to enhance solubilization of digestate as measued by increased levels of soluble Chemical Oxygen Demand (sCOD), soluble Total Organic Carbon (sTOC), and soluble Total Nitrogen (sTN) released into the solution.

Standards for Digestate

The standard of digestate produced by anaerobic digestion can be assessed on three criteria, chemical, biological and physical aspects. Chemical quality needs to be considered in terms of heavy metals and other inorganic contaminant, persistent organic compounds and the content of macro-elements such as nitrogen, phosphorus and potassium. Depending on their source, biowastes can contain pathogens, which can lead to the spreading of human, animal or plant diseases if not appropriately managed.

The physical standards of composts includes mainly appearance and odour factors. Whilst physical contamination does not present a problem with regards to human, plant or animal health, contamination (in the form of plastics, metals and ceramics) can cause a negative public perception. Even if the compost is of high quality and all standards are met, a negative public perception of waste-based composts still exists. The presence of visible contaminants reminds users of this.

Quality control of the feedstock is the most important way of ensuring a quality end product. The content and quality of waste arriving on-site should be characterised as thoroughly as possible prior to being supplied.

In the UK the Publicly Available Specification (called PAS110) governs the definition of digestate derived from the anaerobic digestion of source-segregated biodegradable materials. The specification ensures all digested materials are of consistent quality and fit for purpose. If a biogas plant meets the standard, its digestate will be regarded as having been fully recovered and to have ceased to be waste, and it can be sold with the name "Bio- fertiliser".

PYROLYSIS

Pyrolysis is the chemical decomposition of organic (carbon-based) materials through the application of heat. Pyrolysis, which is also the first step in gasification and combustion, occurs in the absence or near absence of oxygen, and it is thus distinct from combustion (burning), which can take place only if sufficient oxygen is present. The rate of pyrolysis increases with temperature. In industrial applications the temperatures used are often 430 °C (about 800 °F) or higher, whereas in smaller-scale operations the temperature may be much lower. Two well-known products created by pyrolysis are a form of charcoal called biochar, created by heating wood, and coke (which is used as an industrial fuel and a heat shield), created by heating coal. Pyrolysis also produces condensable liquids (or tar) and noncondensable gases.

Process

Pyrolysis transforms organic materials into their gaseous components, a solid residue of carbon and ash, and a liquid called pyrolytic oil (or bio-oil). Pyrolysis has two primary methods for removing contaminants from a substance: destruction and removal. In destruction, the organic contaminants are broken down into compounds with lower molecular weight, whereas in the removal process, they are not destroyed but are separated from the contaminated material. Pyrolysis is a useful process for treating organic materials that "crack" or decompose under the presence of heat; examples include polychlorinated biphenyls (PCBs), dioxins, and polycyclic aromatic hydrocarbons (PAHs). Although pyrolysis is not useful for removing or destroying inorganic materials such as metals, it can be used in techniques that render those materials inert.

Applications

Pyrolysis has numerous applications of interest to green technology. It is useful in extracting materials from goods such as vehicle tires, removing organic contaminants from soils and oily sludges, and creating biofuel from crops and waste products. Pyrolysis can help break down vehicle tires into useful components, thus reducing the environmental burden of discarding the tires. Tires are a significant landfill component in many areas, and they release PAHs and heavy metals into the air when they are burned. However, when tires are pyrolyzed, they break down into gas and oil (usable for fuel) and carbon black (usable as filler in rubber products, including new tires, and as activated charcoal in filters and fuel cells). In addition, pyrolysis can remove organic contaminants, such as synthetic hormones, from sewage sludge (semisolid materials that remain after wastewater is treated and the water content reduced) and make heavy metals remaining in the sludge inert, which allows the sludge to be used safely as fertilizer.

Furthermore, pyrolyzing biomass (biological materials such as wood and sugarcane) holds great promise for producing energy sources that could supplement or replace petroleum-based energy. Pyrolysis causes the cellulose, hemicellulose, and part of the lignin in the biomass to disintegrate to smaller molecules in gaseous form. When cooled, those gases condense to the liquid state and become bio-oil, while the remainder of the original mass (mainly the remaining lignin) is left as solid biochar and noncondensable gases.

Types of Pyrolysis

There are three types of pyrolytic reactions differentiated by the processing time and temperature of the biomass.

Slow Pyrolysis

Slow pyrolysis is characterized by lengthy solids and gas residence times, low temperatures and slow biomass heating rates. In this mode, the heating temperatures ranges from 0.1 to 2 °C (32.18 to 35.6 °F) per second and the prevailing temperatures are nearly 500 °C (932 °F). The residence time of gas may be over five seconds and that of biomass may range from minutes to days.

During slow pyrolysis, tar and char are released as main products as the biomass is slowly devolatilized. Repolymerization/recombination reactions occur after the primary reactions take place.

Flash Pyrolysis

Flash pyrolysis occurs at rapid heating rates and moderate temperatures between 400 and 600 °C (752 and 1112 °F). However, vapor residence time of this process is less than 2s. Flash pyrolysis produces fewer amounts of gas and tar when compared to slow pyrolysis.

Fast Pyrolysis

This process is primarily used to produce bio-oil and gas. During the process, biomass is rapidly heated to temperatures of 650 to 1000 °C (1202 to 1832 °F) depending on the desired amount of bio-oil or gas products. Char is accumulated in large quantities and has to be removed frequently.

Microwave Pyrolysis

Fast pyrolysis has been shown to benefit from the use of microwave heating. Biomass typically absorbs microwave radiation very well, making heating of the material highly efficient - just like microwave heating of food, it can reduce the time taken to initiate the pyrolysis reactions, and also greatly reduces the energy required for the process. Because microwave heating can initiate pyrolysis at much lower overall temperatures (sometimes as low as 200-300°C), it has been found that the bio-oil produced contains higher concentrations of more thermally labile, higher value chemicals, suggesting that microwave bio-oil could be used as a replacement to crude oil as a feedstock for some chemical processes.

Pyrolysis Reactors

Some of the reactors used in pyrolysis process include the following:

Bubbling Fluidized Bed Pyrolyzers

Fluidized beds are generally simple to construct and design when compared to other reactor types. Bubbling fluidized bed pyrolyzers have large heat storage capacity, better temperature control, excellent heat transfer characteristics and better gas-solids contact. In this pyrolyzer, the residence time of vapors and solids are controlled by the fluidizing gas flow rate. During the pyrolysis reaction, char acts as a catalyst in cracking vapors. Char is finally collected by an entrainment processes.

Circulating Fluid Beds and Transported Bed

Circulating fluidized bed pyrolyzers have similar characteristics as that of bubbling bed pyrolyzers excluding that the residence time of vapors and char is faster due to higher gas velocities. These pyrolyzers have better gas-solid contact, high processing capacity and potential to deal with cohesive solids that might otherwise be hard enough to fluidize in bubbling fluidized beds.

Ablative Pyrolyzer

The ablative pyrolyzer, on the other hand, was designed such that the heat transferred from a hot reactor wall tends to soften the feedstock under pressure. Large feedstock particles can be pyrolyzed in this pyrolyzer as the reaction rates are not influenced by heat transfer via the biomass particle. These pyrolyzers ensure high relative motion between the reactor wall and the particle and high pressure of particle on hot reactor wall. It avoids the need of inert gas and hence its processing equipment is small and reaction system is more intense.

Advantages of Pyrolysis

The key benefits of pyrolysis include the following:

- It is a simple, inexpensive technology for processing a wide variety of feedstocks.
- It reduces wastes going to landfill and greenhouse gas emissions.
- It reduces the risk of water pollution.
- It has the potential to reduce the country's dependence on imported energy resources by generating energy from domestic resources.
- Waste management with the help of modern pyrolysis technology is inexpensive than disposal to landfills.
- Construction of a pyrolysis power plant is a relatively rapid process.
- It creates several new jobs for low-income people based on the quantities of waste generated in the region, which in turn provides public health benefits through waste clean up.

Applications

Some of the major applications of pyrolysis include the following:

- It is widely used in chemical industry to produce methanol, activated carbon, charcoal and other substances from wood.
- Synthetic gas produced from the conversion of waste using pyrolysis can be used in gas or steam turbines for producing electricity.
- A mixture of stone, soil, ceramics and glass obtained from pyrolytic waste can be used as a building material - construction slag or for filling landfill cover liners.
- It plays a major role in carbon-14 dating and mass spectrometry.
- It is also used for several cooking procedures like caramelizing, grilling, frying and baking.

BIOREMEDIATION

Bioremediation refers to the use of microorganisms to degrade contaminants that pose environmental and human risks. Bioremediation processes typically involve the actions of many different microbes acting in parallel or sequence to complete the degradation process. Both in situ (in place) and ex situ (removal and treatment in another place) remediation approaches are used. The versatility of microbes to degrade a vast array of pollutants makes bioremediation a technology that can be applied in different soil conditions. Though it can be inexpensive and in situ approaches can reduce disruptive engineering practices, bioremediation is still not a common practice.

A widely used approach to bioremediation involves stimulating naturally occurring microbial communities, providing them with nutrients and other needs, to break down a contaminant. This is termed biostimulation. Biostimulation can be achieved through changes in pH, moisture, aeration, or additions of electron donors, electron acceptors or nutrients. Another bioremediation approach is termed bioaugmentation, where organisms selected for high degradation abilities are used to inoculate the contaminated site. These two approaches are not mutually exclusive- they can be used simultaneously.

Recent awareness of the dangers of many chemicals used in society has led to research on formulation of products that are more easily degraded in the environment.

From an ecological point of view, bioremediation depends on the various interactions between three factors: substrate (pollutant), organisms, and environment, as shown in the figure at right. The interactions of these factors affect biodegradability, bioavailability, and physiological requirements, which are important in assessing the feasibility of bioremediation. Biodegradability, or whether a chemical can be degraded or not, is determined by the presence or absence of organisms that are able to degrade a chemical of interest and how widespread these organisms are in the site. The substrate (pollutant) can interact with its surrounding environment to change its bioavailability, or availability to organisms that are capable of degrading it; for example, substrate has low bioavailability if it is tightly bound to soil organic matter or trapped inside aggregates. Physiological requirements, or set of conditions required by organisms to carry out bioremediation in the environment, include nutrient availability, optimal pH, and availability of electron acceptors, such as oxygen and nitrate. Also, the environment needs to be habitable for organisms involved in bioremediation.

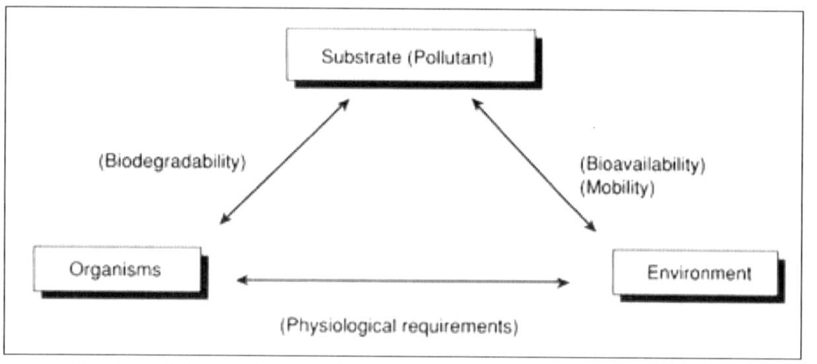

Retrieved from Bioremediation from an ecological perspective.

Microorganisms in the environment have always broken down waste, and humans have always (knowingly or unknowingly) used them in agricultural, domestic, and industrial activities. As the urbanized world shifted to a more industrial system, however, people began to take an active approach in bioremediation. In the late nineteenth century, wastewater treatment plants were formed, but even so, this was not officially called bioremediation. The project considered the initial spark of the bioremediation movement was the report "Beneficial Stimulation of Bacterial Activity in Groundwater Containing Petroleum Products" by R.L. Raymond et al. in 1975. By testing the relationship between oil presence and bacterial stimulation, Raymond found that adding nutrients to soil hastened the oil removal. This led to the development of in situ bioremediation.

Initial bioremediation projects focused on "pump and treat" (ex situ) methods in soil around gas stations and refinery spills to get oil out of groundwater sources, but soon cleaning up chlorinated hydrocarbons became a primary concern. Chlorinated compounds were commonly used in pesticides, but when people learned it was a possible carcinogen and causing ozone depletion, research into bioremediation took off. This was when anaerobic bacteria started being used, as it was discovered that they dechlorinate compounds much more quickly than do aerobic bacteria, and produce fewer damaging iron compounds that precipitate from the reactions.

First Water Treatment Facility in Japan.

Pollutants

Pollutants found in soils present a variety of different human health risks. Soil pollutants are typically classified as organic and inorganic pollutants.

Organic Pollutants

Industrialization resulted in increased use of organic compounds that build up and persist in the environment. Main sources of organic pollutants are through anthropogenic activities, including use of solvents, pesticides, and fuels. Some of these organic compounds are highly toxic and they are associated with variety of health issues around the world.

While organic pollutants are causing both environmental and health problems, bioremediation offers an effective solution to the pollution.

Inorganic Pollutants

Table: Most inorganic pollutants are due to human activities.

Pollutant	Source
Arsenic	Pesticides, wood preservatives, biosolids, ore mining and smelting.
Cadmium	Paints and pigments, plastic stabilizers, electroplating, phosphate fertilizers.
Chromium	Tanneries, steel industries, fly ash.
Copper	Pesticides, fertilizers, biosolids, ore mining and smelting.
Mercury	Gold and Silver mining, coal combustion.
Nickel	Effluent, kitchen appliances, surgical instruments, automobile batteries.
Lead	Aerial emission from combustion of leaded fuel, batteries waste, insecticide and herbicides.

A majority of heavy metal pollutants come from human sources that accumulate over time. There are also natural forms of contamination from normal biological processes, which include:

- Weathering of minerals over time.
- Erosion and volcanic activities.
- Forest fires and biogenic source.
- Particles released by vegetation.

Heavy metals can be absorbed by microbes at cellular binding sites. Extracellular polymers of these microbes can complex heavy metals through various mechanisms. These specialized microorganisms can mineralize the organic contaminants to metabolic intermediates, which are used as primary substrates for cell growth. The microbes prevalent in heavily metal-contaminated soil can alter the oxidation state of the heavy metals by immobilizing them, allowing them to be easily removed. Bioremediation of heavy metals from microbes is not heavily researched, mostly due to an incomplete understanding of the genetics of the microbes used in metal adsorption. Geomicrobiology takes a better look at the interactions between microbes and inorganic material.

Organisms

As stated previously, bioremediation involves various microorganisms that are able to degrade and reduce toxicity of environmental pollutants. Therefore, the interactions of microbes with the environment and pollutants are significant in determining effectiveness of bioremediation. Those microbes can be either naturally present in the site of bioremediation or isolated from other sites and inoculated artificially. Biodegradation often occurs as part of microbial metabolism and in some cases, microbes are able to directly harvest carbon and energy by breaking down pollutants. Sections below go over bacteria and fungi, the commonly used organisms in bioremediation, and archaea, the more recently discovered group of organisms with unique potential in bioremediation.

Bacteria

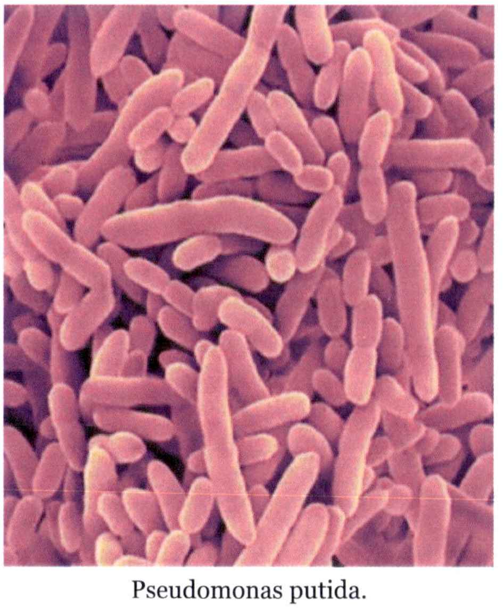

Pseudomonas putida.

Bacteria are widely diverse organisms, and thus make excellent players in biodegradation and bioremediation. There are few universal toxins to bacteria, so there is likely an organism able to break down any given substrate, when provided with the right conditions (anaerobic versus aerobic environment, sufficient electron donors or acceptors, etc.). Below are several specific bacteria species known to participate in bioremediation.

Pseudomonas Putida

Pseudomonas putida is a gram-negative soil bacterium that is involved in the bioremediation of toluene, a component of paint thinner. It is also capable of degrading naphthalene, a product of petroleum refining, in contaminated soils.

Dechloromonas Aromatica

Dechloromonas aromatica is a rod-shaped bacterium which can oxidize aromatics including benzoate, chlorobenzoate, and toluene, coupling the reaction with the reduction of oxygen, chlorate, or nitrate. It is the only organism able to oxidize benzene anaerobically. Due to the high propensity of benzene contamination, especially in ground and surface water, D. aromatic is especially useful for in situ bioremediation of this substance.

Nitrifiers and Denitrifiers

Industrial bioremediation is used to clean wastewater. Most treatment systems rely on microbial activity to remove unwanted mineral nitrogen compounds (i.e. ammonia, nitrite, nitrate). The removal of nitrogen is a two stage process that involves nitrification and denitrification. During nitrification, ammonium is oxidized to nitrite by organisms like Nitrosomonas europaea.Then, nitrite is further oxidized to nitrate by microbes like Nitrobacter hamburgensis.

In anaerobic conditions, nitrate produced during ammonium oxidation is used as a terminal electron acceptor by microbes like Paracoccus denitrificans. The result is N2 gas. Through this process, ammonium and nitrate, two pollutants responsible for eutrophication in natural waters, are remediated.

Deinococcus Radiodurans

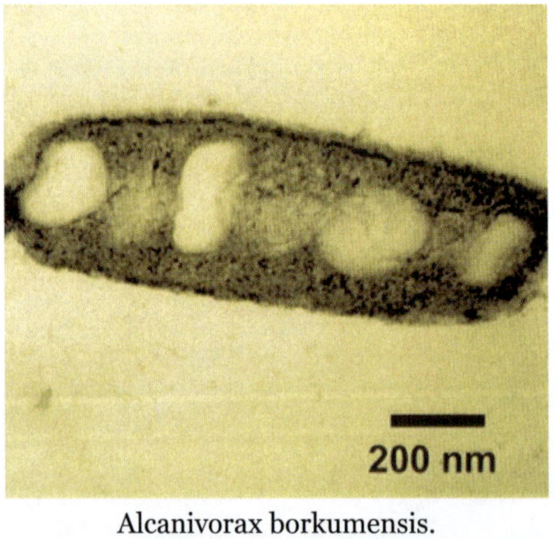

Alcanivorax borkumensis.

Deinococcus radiodurans is a radiation-resistant extremophile bacterium that is genetically engineered for the bioremediation of solvents and heavy metals. An engineered strain of Deinococcus radiodurans has been shown to degrade ionic mercury and toluene in radioactive mixed waste environments.

In anaerobic conditions, nitrate produced during ammonium oxidation is used as a terminal electron acceptor by microbes like Paracoccus denitrificans. The result is dinitrogen gas. Through this process, ammonium and nitrate, two pollutants responsible for eutrophication in natural waters, are remediated.

Methylibium Petroleiphilum

Methylibium petroleiphilum (formally known as PM1 strain) is a bacterium capable of methyl tert-butyl ether (MTBE) bioremediation. PM1 degrades MTBE by using the contaminant as the sole carbon and energy source.

Alcanivorax Borkumensis

Alcanivorax borkumensis is a marine rod-shaped bacterium which consumes hydrocarbons, such as the ones found in fuel, and produces carbon dioxide. It grows rapidly in environments damaged by oil, and has been used to aid in cleaning the more than 830,000 gallons of oil from the Deepwater Horizon oil spill in the Gulf of Mexico.

Fungi (Mycoremediation)

Current bioremediation applications primarily utilize bacteria, with comparatively few attempts to use fungi. Fungi have fundamentally important roles because of their participation in the cycling of elements through decomposition and transformation of organic and inorganic materials. These characteristics can be translated into applications for bioremediation which could break down organic compounds and reduce the risks of metals. In some cases, fungi have an advantage over bacteria not just in metabolic versatility but also their environmental resilience.They are able to oxidize a diverse amount of chemicals and survive in harsh environmental conditions such as low moisture and high concentrations of pollutants. Therefore, fungi are potentially an extremely powerful tool in soil bioremediation and some versatile species such as White Rot Fungi have been a hot topic of research.

Biodegradation Capacities of White Rot Fungi

Using fungi as potential treatment of contaminants began in 1985 when the white rot species Phanerochaete chrysosporium was discovered to metabolize multiple key environmental pollutants. The most important feature of these fungi is their enzymatic functional ability to metabolize complex chemicals such as lignin. Similar abilities were later discovered in other white rot fungal species. In addition, white rot fungi are highly advantageous because they degrade lignin extracellularly through its hyphal extension. This allows them to access soil contaminants that other organisms are incapable of and maximize surface area for enzymatic interaction. These inexpensive fungi can tolerate extreme environmental conditions, such as pH, temperature, and moisture content. While many microbial organisms that are used for bioremediation require pre-conditioning of the environment for them to survive in, white rot fungi can directly be applied into most systems because they degrade based upon nutrient deprivation.

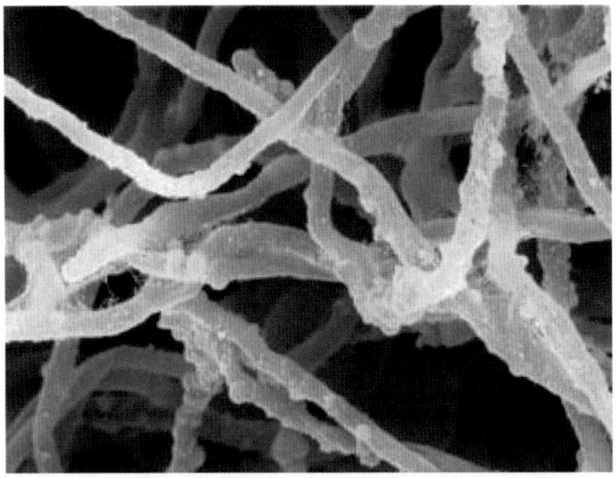

Scanning electron micrograph (SEM) depicts Phanerochaete chrysosporium fungi.

Phanerochaete Chrysosporium

P. chrysosporium was the first fungi linked to degradation of organic pollutants. Extensive research has show this it has strong potential for bioremediation in pesticides, PAHs, dioxins, carbon tetrachloride, and many other pollutants. Among fungal systems, P. chrysosporium has become the model for bioremediation. Other notable species of white rot fungi include Pleurotus ostreatus and Trametes versicolor.

Bioremediation of Hydrocarbon Pollutants

Hydrocarbons are stored deep underground but are brought up to the surface to be transformed and utilized, primarily as an energy source known as fossil fuels. The majority of pollution currently comes from these byproducts in the form Polycyclic Aromatic Hydrocarbons (PAHs), which are xenobiotic environmental pollutants that form when carbon materials are incompletely combusted. Some of examples of PAHs include burning wood, fossil fuels, and cigarette smoke. [19,20] Currently, bioremediation is only effective for soils contaminated with low-molecular weight PAHs because of bacterial commercial use. However, fungi are effective at PAH degradation in comparison to bacteria for a few reasons. Firstly, they are capable degrading PAH's that are high in molecular weight, bacteria in comparison are better at degrading smaller molecules. Secondly, fungi can function well in non-aqueous environments and low oxygen conditions, both are conditions where PAH's can accumulate. Many fungi have evolved mechanisms that allow the to target specific PAHs. Fungi produce extracellular enzymes that degrade lignin, a process called mineralization the produces carbon dioxide as the end product.

Remediating Metals

Toxic metals can enter the environment all life cycle stages of metal compound. For example, metal leaching can occur from the mining process till the disposal of metal wastes. However in nature, the mobility of metals comes from the geological processes that can be released into the soil and aquatic environments. The environmental largest risk from metal contamination comes from the relationship between metals and compounds that are inherently of incapable of being degraded by any natural procedures. The best solution to treating contamination is transporting the metals to location where they cannot produce negative environmental effects. Fungi have various ways of interacting with metals, some of the techniques are increasing or decreasing the mobility of metals, sorption, or even cellular uptake. After the metals have been absorbed the fungus, they can chemically altered to be stored or translocated through the hyphae and into various plants that participate in symbiosis.

Pesticide Degradation

Pesticide accumulation is an issue of great concern among the public, because they are directly associated with food products and water supplies. There are number of technologies used for pesticide clean-up; however, these technologies are generally expensive and inefficient because they require contaminated soil to be excavated and sent to a separate storage location for processing. Bioremediation offers a potential solution that treats contaminated soil and groundwater without needing excavation. Studies show that White Rot Fungi has high promise for soil bioremediation application; however, most tests have been conducted in the lab rather than in the actual environment. This fungi demonstrates the ability to transform and mineralize specific pesticides in soil.

Environmental Applications

Although fungi demonstrate significant biochemical and ecological useful qualities, they are hardly utilized for biotechnological purposes. Instead, bacteria are most commonly used because they usually produce superior results in their numerous advantages ranging from their highly specific biochemical reactions to their capabilities of breaking down pollutants efficiently. Fungi are underused primarily because of the costs that come from providing oxygen to fungi in polluted environments. However, filamentous fungi could be highly valuable in situations where bacteria cannot perform. For example, fungi are useful in situations where contaminants are physically blockaded and bacteria cannot reach or in circumstances of environmental extremes such as high acidity or dryness prevent bacteria from functioning.

Archaea

The role of archaea in bioremediation has not been studied as commonly as that of bacteria. Nevertheless, numbers of researchers have shown their ability to degrade various pollutants and scientists began to discover more about their potential in participating in bioremediation. Below lists some important facts regarding archaea's potential role in bioremediation:

- Biodegradation by extreme halophilic archaea was not recognized widely in the past, but scientists have found out that extreme halophilic archaea have greater catabolic diversity than expected.
- Hydrocarbon-contamination is observed in some extreme environments, including hypersaline (high salt concentration), high or low temperature, or extreme pH. Archaea's adaptation to extreme environment gives them the potential to participate in biodegradation and bioremediation in these environments; in fact, microorganisms naturally adapted to the cold environments are known to be important degraders of hydrocarbons in those environments.

- Extreme halophilic archaea has potential to biodegrade pollutants in hypersaline environment, in which bacteria typically used in bioremediation cannot survive or function properly.
- Some archaea are known to be resistant to variety of antibiotics, including penicillin, cycloheximide, streptomycin, etc, which gives them great advantage in participating in bioremediation in the presence of antibiotics.

Examples of Studies of Archaea Involved in Bioremediation

Four extreme halophilic strains of archaea (belonging to genus Halobacterium, Haloferax, and Halococcus) were studied to evaluate their potential to biodegrade crude oil and hydrocarbons. All four strains could use various kinds of hydrocarbons as their carbon or energy sources. Two strains of Haloferax grew on n-alkanes with different lengths, ranging from C8 to C34, and also benzene, toluene, biphenyl, and naphthalene. The research demonstrated the important fact that archaea have potential to carry out biodegradation at high temperatures, in the range of 40-45 °C, which is advantageous because hydrocarbons have higher solubility and bioavailability at these higher temperature. The four strains studied were resistant to six different antibiotics, including penicillin, streptomycin, cycloheximide and this gave them the potential to carry out biodegradation in conditions unfavorable for bacteria. Research suggests other genera of archaea are also capable of biodegrading in hypersaline environments.

Archaeglobus fulgidus, a hyperthermophile which can use sulfate as an electron acceptor, can also break down various aromatic hydrocarbons.

Microbial Processes

Microorganisms use a wide range of processes to transform chemicals in their environment. In some cases, pollutants serve as the carbon and energy source for microbial growth, while in other cases, pollutants serve as the terminal electron acceptor. This manifests itself in the diverse ability of microbes to transform and degrade toxic molecules. Below, several steps and details of the microorganisms' actions are described.

Factors affecting Rates of Biodegradation

Biodegradation may be influenced by pH, temperature, moisture, carbon sources, soil texture, aerobic versus anaerobic conditions, the number of substituents, and the concentration of the pollutant. It is impossible, however, to make a generalization about the best universal conditions for biodegradation. What's toxic to some microbes is a nutrient to others, what might be a damaging pH to some is beneficial to others, and so on.

A greater amount of substituents will cause slower degradation in aerobic environments, but faster degradation in anaerobic ones. Chlorine makes a molecule less degradable due to steric hindrance preventing access to necessary enzymes, therefore molecules with higher chlorination are slower to degrade in aerobic conditions. High concentration of a pollutant generally results in faster rates of degradation. If the concentration drops below a threshold concentration, the enzymes may not detect it and will cease to degrade it.

Soil with small pores, especially clays, may cause biodegradation to take years due to the decrease in bioavailability. Chlorine makes a molecule less degradable due to steric hindrance preventing necessary enzymes from accessing the compound, therefore molecules with higher chlorination are slower to degrade.

The rate at which a compound is transformed, as well as the curves that describe its transformation, is referred to as kinetics, and is affected by all factors listed above. First order kinetics (exponential decay) often describes biodegradation when the initial substrate concentration is low, while zero-order kinetics (linear biodegradation) is often observed when the substrate concentration is very high. In some cases if the concentration of the chemical falls below a critical threshold concentration, the microbes can no longer transform it and the chemical persists.

The power rate model depicting the relationship between concentration and rate of degradation (first order decay here) is as follows:

$$-dC/dt = kC\text{^}n$$

C is substrate concentration, t is time, k is a rate constant for the chemical in question, and n is an appropriate parameter. The values of k and n are adjusted until a line is found to match experimental data.

Primary Substrate Utilization

Primary substrate utilization occurs when a microbe both transforms a substrate and uses it as an energy or carbon source. An electron acceptor is required for these transformations. It can be anaerobic or aerobic, although the presence of oxygen tends to speed up reactions. This type of biodegradation is common in break down of petroleum compounds and some pesticides.

Cometabolism (Secondary Substrate Utilization)

Cometabolism involves the fortuitous transformation of a chemical by an organism while the organism uses a different substance as its primary energy or carbon source. During the actual reaction degrading the substance, it appears that the organism involved has no net carbon or energy gain, and may even result in a product which is toxic to the cell.

A key example of cometabolism is fortuitous metabolism in the degradation of trichloroethylene, shown in the diagram below. An organic growth substrate such as propane or butane is required for the enzymatic activity that transforms TCE.

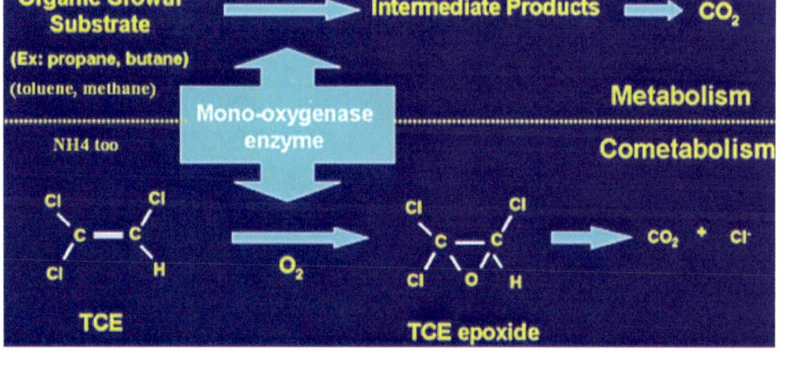

Reductive and Hydrolytic Dehalogenation

Chloride and other halogens are common components of pesticides and hazardous industrial wastes, and by removing them the toxic chemical can often be remediated. If the halogen is replaced by a hydrogen (RCl -> RH), then it is reductive dehalogenation. If two halogens are replaced simultaneously, then the process is called dihaloelimination, although it still falls under reductive dehalogenation. If the halogen is replaced by OH (RCl -> ROH) then it's hydrolytic dehalogenation. In both cases, the halogen is released as its inorganic form into the environment.

Acclimation

An acclimation period, also called an adaptation or lag period, occurs when no destruction of a given chemical is observed. It is caused by the microbes transitioning to their altered environment and shifting their metabolism to better suit it. It can last for anywhere from hours (such as aromatic compounds in warm, oxygenated soils) to months (such as halobenzoates in anaerobic sediments) depending on the chemical in question and the environment. Acclimation periods can be affected by temperature, the presence of oxygen, pH, and concentration of the substance. Although they are most often faster in warm, aerated, and fairly dry environments, there are few consistencies between what shortens or lengthens the period, even if the concentration is the same. Insecticides including methyl parathion and azinphosmethyl; herbicides including 2, 4-D, MCPA, Mecoprop, TCA, and amitrole; the quaternary ammonium compound dodecyltrimethylammonium chloride; polycyclic aromatic hydrocarbons including naphthalene and anthracene; and other chemicals such as phenol, chlorobenzene, PCP, diphenyl-methane, and NTA have all been reported to have acclimation periods, and this can be of severe human concern. The continued presence of these toxins extends human, plant, and animal exposure, and if the chemical is in water, it can allow the substance to flow further and impact environments distant to its site of origin before being degraded.

Detoxification and Activation

Detoxication, sometimes called detoxification, has been referred to as the "most important role of microorganisms in the transformation of pollutants". The process is the changing of a molecule into something less harmful to a species in question. There are a number of ways a molecule can be transformed, including hydrolysis, hydroxylation, dehalogenation, demethylation, methylation, and ether cleavage. By breaking bonds, or adding or removing groups, the organism reduces its effect on the environment. Furthermore, although sometimes the resulting chemical is simply excreted as waste, the organism may also be able to use this new compound as a carbon source or further modifies it until it is released as CO_2.

There are instances where the initial compound is harmless, and in fact the substance produced by microorganisms, or an intermediate in the degradation process, is a toxin. This process is called activation. For this reason, it is important to test all steps of a reaction when determining how a compound is degrading. The new toxins may also be more or less mobile than its predecessor, so it can either stick around one area for extended periods of time or spread to other areas and increase damage. A prevalent example of this is the dechlorination of TCE, which produces DCE (50 times more hazardous than TCE) and Vinyl Chloride (a known carcinogen). Commonly used insecticides in the past, like zinophos, trichloronat, and carbofuran, were all found to increase a soil's toxicity with extended use.

Bioremediation Treatment Methods

In order for bioremediation to be successful, it requires sufficient proof for the degradation of contaminants. However, determining the effectiveness and completeness to reach sufficient results is one of the major issues. Natural attenuation relies on natural processes to clean up or attenuate pollution in soil and groundwater. This remediation is done without human interaction, and is primarily used as a monitoring technique, to make sure more aggressive cleanup strategies are not needed. Abiotic and biotic factors play a distinguishing factor of how effective bioremediation is.

Current monitoring practices determine the disappearance of contaminants and their degradation products to regulatory levels that are monitored by toxicity testing, usually on single organisms or species to ensure there are no induced changes that may result in residual toxicity. The problem with these monitoring techniques is that the assessment of contaminants may result in an inaccurate indicator of residual toxicity. Rather, studying the microbial community response may be a more comprehensive indicator of residual toxicity than a single species. Once sufficient evidence is provided, human intervention may be needed for a more effective cleanup process.

There are two types of remediation that are done, ex situ: which is done by removing the contaminated soil or water and treating it outside the source, and in situ: which treatment takes place within the contaminated area. There are some treatments methods that can be either ex situ or in situ. Some techniques may deal with the mobilization of pollutants, to move them out of an area, or immobilized to keep them out of an area such as a water table.

Table: A comparative analysis of the different types of bioremediation. It can be used to find which remediation technique may be used in certain circumstances.

Technology	Examples	Benefits	Limitations	Factors to consider
In situ	In situ bioremediation Biosparging Bioventing Bioaugmentation	Most cost efficient Noninvasive Relatively passive Natural allenuation processes Treats soil and water	Environmental constraints Extended treatment time monitoring difficulties	Biodegradative abilities of indigenous microorganisms Presence of metal and other inorganics Environmental parameters Biodegradability of pollutants Chemical solubility Geological factors Distribution of pollutants
Ex situ	Land farming Composting Biopiles	Costs efficient Low cost Can be done on site	Space requirements Extended treatment time Need to control abiotic loss Mass transfer problem Bioavailability limitation	See above
Bioreactors	Slurry reactors Aqueous reactors	Rapid degradation kinetic Optimized environmental parameters Enhances mass transfer Effective use of inoculants and surfactants	Soil requires excavation Relatively high cost capital Relatively high operating cost	See above Bioaugmentation Toxicity of amendment Toxic concentrations of contaminants

Contaminated soil is mixed with amendments and piled on top of a liner, while a pipe with a blower controls aeration.

Ex-situ

Ex-situ techniques are those that are applied to soil and groundwater which has been removed from the site via excavation or pumping. The methods used include composting, biofilters, and biopiling. Ex-situ is used for smaller projects, primarily because larger excavation of soil is not prefered. The movement of the soil can be more detrimental by destroying the preestablish horizons in the soil.

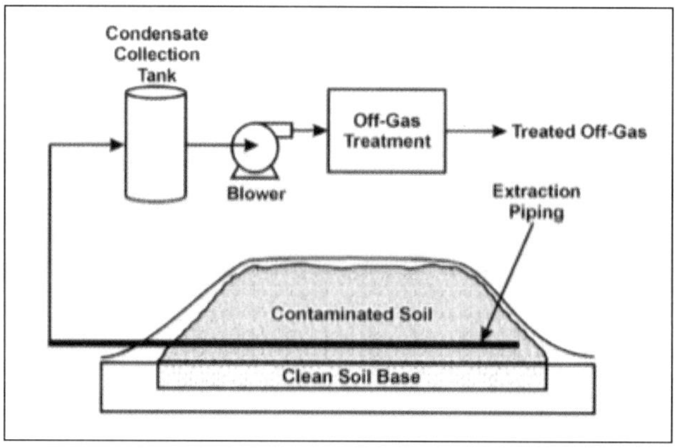

Contaminated soil is mixed with amendments and piled on top of a liner, while a pipe with a blower controls aeration.

Biopiling

Excavated soils are mixed with soil amendments and placed on a treatment area. Biopiles are aerated with the use of perforated pipes and blowers in order to control the progression of biodegradation more efficiently by controlling the supply of oxygen, which in turn may affect other factors such as pH. This system is primarily used to remediate systems with oil and hydrocarbon contamination. The remediated soil is placed in a liner to prevent further contamination of the soil, they may also be covered with plastic to control runoff, evaporation, and volatilization.

Composting

Nutrients are added to soil that is mixed to increase aeration and activation of indigenous microorganisms. Composting is done in a separate container, then when composting is complete it is incorporated into the soil. Bioremediation by the utilization of compost relies on the adsorption capabilities of organic matter and the degradation capabilities of microorganisms present. Composting is recognized as as one of the most cost-effective technologies for soil bioremediation and it can be done on large and small scales. The use of composting is a very versatile technique for soil polluted by a wide range of organic pollutants and heavy metals, making it great for easier remediation involving various pollutants. The utilization of organic wastes for soil remediation is also helpful in decreasing the need for their storage and treatment. Organic matter that is generated from composting offers the benefit of improving soil quality and structure. Composting is primarily used for remediation over a longer period of time, as the nutrients for the microbes are released gradually and requrire more time compared to quicker treatments such as biostimulation.

In-situ

In-situ techniques are applied to soil and groundwater at the site with minimal disturbance. These methods include biostimulation, bioleaching, biosorption, and bioventing. In-situ is preferred because it is often minimally invasive to the soil structure in comparison to ex-situ, but it can be expensive due to specialized equipment.

Biostimulation

This method involves the addition of nutrients to a polluted site in order to encourage the growth of naturally occurring chemical-degrading microorganisms. Biostimulation is primarily done by the addition of various nutrients that are limited in the soil as well as electron acceptors, such as phosphorus, nitrogen and oxygen, or increasing the amount of available carbon in order to increase the population or activity of naturally occurring microorganisms. Other approaches are to optimize environmental conditions such as aeration, the addition of nutrients, altering pH and temperature control. The primary advantage of biostimulation is that it is done by native microorganisms that are well-suited to the environment, and are already well distributed spatially. The challenge is delivering additives so they are readily available to the subsurface microbes.

Metal Biosorption

Adsorption of metals and other ions of an aqueous solution by the use of microbes. The biosorption process involves a solid phase and a liquid phase containing a dissolved species to be sorbed. The process continues until equilibrium is established between the amount of solid-bound sorbate species and its portion remaining in the solution. The degree of affinity for the sorbate determines its distribution between the solid and liquid phases.

Biosorption processes are very important in the environment, and has been utalized for conventional biotreatment processes. Biosorption is primarily aimed at the removal or recovery of organic and inorganic substances from solution. The commercialization of biosorption technologies has been limited so far.

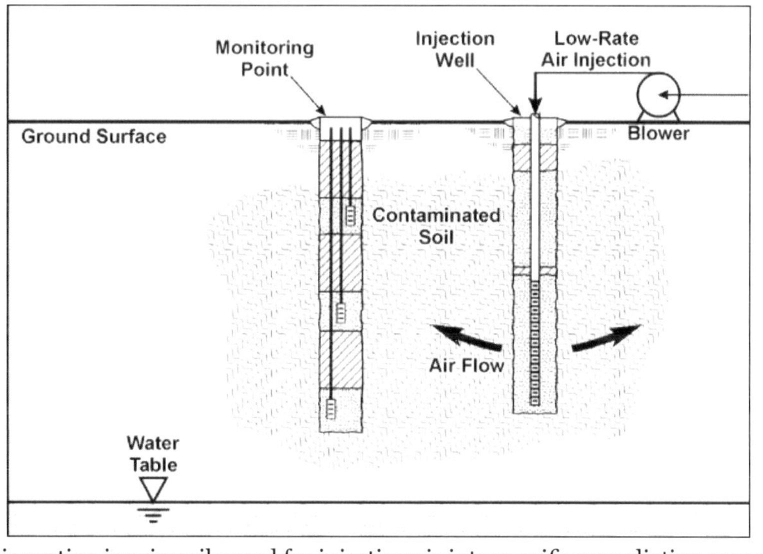

Bioventing is primarily used for injecting air into specific remediation zones.

Bioventing

Bioventing is an In situ remediation technology that uses indigenous microorganisms to biodegrade organic constituents adsorbed to soils in the unsaturated zone. The availability of oxygen generally controls the rate at which aerobic bioremediation proceeds. Bioventing is the coupling of soil venting and bioremediation. Bioventing can be successfully applied to compounds ranging from gasoline or diesel, to heavier hydrocarbons. The addition of nutrients with the bioventing flow rates can achieve greater contaminant reductions than venting alone.

Ex-situ or In-situ

Some methods can be used by either in-situ or ex-situ methods. The soil or water can be removed from the contamination source and treated, or treated at the source, the method chosen can be based on many factors such as how expensive the project may be or how much contaminant needs to be treated. These methods include bioaugmentation, land farming and biofiltration.

Bioaugmentation

Bioaugmentation is the addition of non-native microorganisms that have the ability to degrade the contaminants that are recalcitrant to the indigenous microbiota. Bioaugmentation has been proven successful in cleaning organic pollutant, but still faces many environmental problems, such as the survival of strains introduced to soil. The number of introduced microorganisms usually decreases shortly after soil inoculation.

Bioaugmentation is ideal for soil:

- With low number of microbes that are capable of degrading targeted pollutants.
- Containing compounds requiring multi stepped remediation.

Augmentation techniques have a great potential for aromatic compound remediation. The most important step in successful bioaugmentation is selection of proper microbial strains. The success of bioaugmentation strongly depends on the ability of inoculants to survive in contaminated soil, which may vary due to predation and an environment that does provide all the conditions and nutrients that the organism needs to survive. In some cases the environment may be toxic to the added organism.

Land Farming

Contaminated soil is mixed with amendments such as nutrients, and then they are tilled into the earth, or the contaminated soil is applied into lined beds and periodically turned over or tilled to aerate the waste. The topmost layer is the area of concentration for this method, so it is not ideal for deeper remediation. Land farming differs from composting because it actually incorporates contaminated soil into soil that is uncontaminated. The higher zone of remediation will typically contain primarily lighter hydrocarbons that can be volatilized. The material is periodically tilled for aeration to hasten remediation of any nutrients and allow more oxygen to act as electron acceptors, as well as allowing volatilization to occur. Contaminants are degraded, transformed, and immobilized by microbiological processes and oxidation. Soil conditions are controlled to optimize the rate of contaminant degradation, moisture content, frequency of aeration, and pH are all conditions that may be controlled.

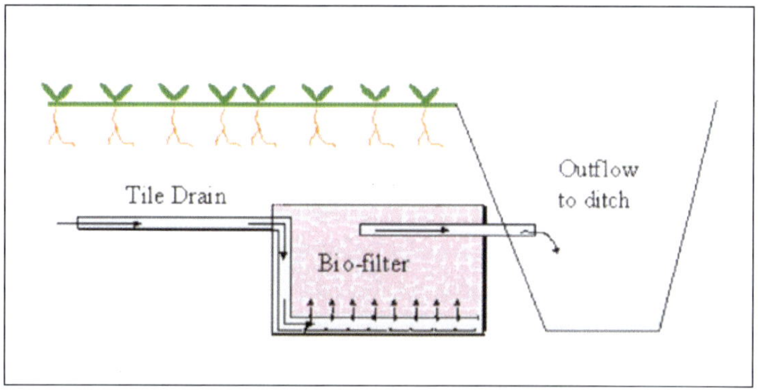

The application of a micro-algal/bacterial biofilter in the primary outflow of soil water.

Biofilter

Biofilters are primarily used for the filtration of contaminated groundwater in the soil. Biofilters can be used above soil, where the water will be pumped aboveground for treatment, or a filter can be placed in the soil near an outflow. A micro-algal/bacterial biofilter can be used for the detoxification of copper and cadmium metal wastes. Biofilters have been used in larger industry environments to treat contaminated outflow of water. Chromobacterium violaceum, is used to treat water and soil contaminated with silver nanoparticles, reducing its concentration.

Bioremediation Synopsis

Advantages

- Bioremediation that involves natural attenuation or biostimulation is a publicly accepted treatment of polluted soil because it is based upon natural processes. Microbes that metabolize contaminants often increase in population when the contaminant is present and thus rates of biodegradation may increase over time, up to a point. If biodegradation is complete (i.e. mineralization) the products from treatment are harmless; such as carbon dioxide, water, and cellular biomass.
- In situ bioremediation can result in complete degradation of pollutants into harmless products on site. This removes the risks involved with transportation for treatment and elimination of contaminated substances.
- Bioremediation can be a cheaper alternative to other technologies used for pollution mitigation.

Disadvantages

- Only biodegradable compounds are capable of undergoing bioremediation. Not every compound is capable of fully degrading quickly.
- The products of biodegradation may potentially be even more persistent or toxic than the original contaminant.

- Biological functions are usually extremely specific and require the presence of microbes that are capable of metabolizing the contaminants. In order for the correct microbes to be present, the appropriate environmental conditions, levels of nutrients, and contaminants need to be met.
- Scaling up the size of studies from small initial studies to commercial-scale field operations is difficult.
- The real environment contains contaminants that are mixed, unevenly distributed, and in different phases (solid, liquid, gas). More research needs to be completed to create technologies that can adapt.
- Compared to other treatment technologies, bioremediation often takes more time.
- Problems with ensuring adequate contact between the microbes and the contaminant. preferential pathway and soil structure can leave uncertainty in remediation dispersal.

BIODEGRADATION

Biodegradation is the breakdown of organic matter by microorganisms, such as bacteria and fungi.

Mechanisms

The process of biodegradation can be divided into three stages: biodeterioration, biofragmentation, and assimilation. Biodeterioration is sometimes described as a surface-level degradation that modifies the mechanical, physical, and chemical properties of the material. This stage occurs when the material is exposed to abiotic factors in the outdoor environment and allows for further degradation by weakening the material's structure. Some abiotic factors that influence these initial changes are compression (mechanical), light, temperature, and chemicals in the environment. While biodeterioration typically occurs as the first stage of biodegradation, it can in some cases be parallel to biofragmentation. Hueck, however, defined Biodeterioration as the undesirable action of living organisms on Man's materials, involving such things as breakdown of stone facades of buildings, corrosion of metals by microorganisms, or merely the esthetic changes induced on man-made structures by the growth of living organisms.

Biofragmentation of a polymer is the lytic process in which bonds within a polymer are cleaved, generating oligomers and monomers in its place. The steps taken to fragment these materials also differ based on the presence of oxygen in the system. The breakdown of materials by microorganisms when oxygen is present, it's aerobic digestion. And the breakdown of materials when oxygen is not present, is anaerobic digestion. The main difference between these processes is that anaerobic reactions produce methane, while aerobic reactions do not (however, both reactions produce carbon dioxide, water, some type of residue, and a new biomass). In addition, aerobic digestion typically occurs more rapidly than anaerobic digestion, while anaerobic digestion does a better job reducing the volume and mass of the material. Due to anaerobic digestion's ability to reduce the volume and mass of waste materials and produce a natural gas, anaerobic digestion technology is widely used for waste management systems and as a source of local, renewable energy.

The resulting products from biofragmentation are then integrated into microbial cells, this is the assimilation stage. Some of the products from fragmentation are easily transported within the cell by membrane carriers. However, others still have to undergo biotransformation reactions to yield products that can then be transported inside the cell. Once inside the cell, the products enter catabolic pathways that either lead to the production of adenosine triphosphate (ATP) or elements of the cells structure.

$$C_{polymer} + O_2 \longrightarrow C_{residue} + C_{biomass} + CO_2 + H_2O$$

Aerobic biodegradation formula

$$C_{polymer} \longrightarrow C_{residue} + C_{biomass} + CO_2 + CH_4 + H_2O$$

Anaerobic degradation formula

Factors affecting Biodegradation Rate

In practice, almost all chemical compounds and materials are subject to biodegradation processes. The significance, however, is in the relative rates of such processes, such as days, weeks, years or centuries. A number of factors determine the rate at which this degradation of organic compounds occurs. Factors include light, water, oxygen and temperature. The degradation rate of many organic compounds is limited by their bioavailability, which is the rate at which a substance is absorbed into a system or made available at the site of physiological activity, as compounds must be released into solution before organisms can degrade them.The rate of biodegradation can be measured in a number of ways. Respirometry tests can be used for aerobic microbes. First one places a solid waste sample in a container with microorganisms and soil, and then aerates the mixture. Over the course of several days, microorganisms digest the sample bit by bit and produce carbon dioxide – the resulting amount of CO_2 serves as an indicator of degradation. Biodegradability can also be measured by anaerobic microbes and the amount of methane or alloy that they are able to produce.

It's important to note factors that effect biodegradation rates during product testing to ensure that the results produced are accurate and reliable. Several materials will test as being biodegradable under optimal conditions in a lab for approval but these results may not reflect real world outcomes where factors are more variable. For example, a material may have tested as biodegrading at a high rate in the lab may not degrade at a high rate in a landfill because landfills often lack light, water, and microbial activity that are necessary for degradation to occur. Thus, it is very important that there are standards for plastic biodegradable products, which have a large impact on the environment. The development and use of accurate standard test methods can help ensure that all plastics that are being produced and commercialized will actually biodegrade in natural environments. One test that has been developed for this purpose is DINV 54900.

Approximated time for compounds to biodegrade in a marine environment	
Product	Time to Biodegrade
Paper towel	2–4 weeks
Newspaper	6 weeks
Apple core	2 months
Cardboard box	2 months

Wax coated milk carton	3 months
Cotton gloves	1–5 months
Wool gloves	1 year
Plywood	1–3 years
Painted wooden sticks	13 years
Plastic bags	10–20 years
Tin cans	50 years
Disposable diapers	50–100 years
Plastic bottle	100 years
Aluminium cans	200 years
Glass bottles	Undetermined

Time-frame for common items to break down in a terrestrial environment	
Vegetables	5 days – 1 month
Paper	2–5 months
Cotton T-shirt	6 months
Orange peels	6 months
Tree leaves	1 year
Wool socks	1–5 years
Plastic-coated paper milk cartons	5 years
Leather shoes	25–40 years
Nylon fabric	30–40 years
Tin cans	50–100 years
Aluminium cans	80–100 years
Glass bottles	1 million years
Styrofoam cup	500 years to forever
Plastic bags	500 years to forever

Plastics

The term Biodegradable Plastics refers to a material that maintains its mechanical strength during practical use but break down into low-weight compounds and non-toxic byproducts after their use. This breakdown is made possible through an attack of microorganisms on the material, which is typically a non-water soluble polymer. Such materials can be obtained through chemical synthesis, fermentation by microorganisms, and from chemically modified natural products.

Plastics biodegrade at highly variable rates. PVC-based plumbing is selected for handling sewage because PVC resists biodegradation. Some packaging materials on the other hand are being developed that would degrade readily upon exposure to the environment. Examples of synthetic polymers that biodegrade quickly include polycaprolactone, other polyesters and aromatic-aliphatic esters, due to their ester bonds being susceptible to attack by water. A prominent example is poly-3-hydroxybutyrate, the renewably derived polylactic acid, and the synthetic polycaprolactone. Others are the cellulose-based cellulose acetate and celluloid (cellulose nitrate).

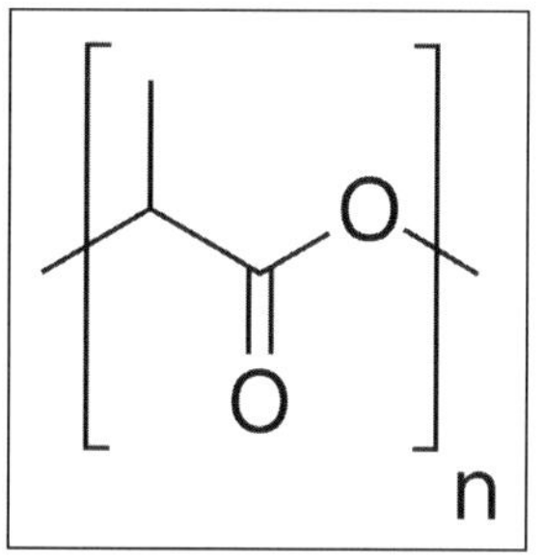

Polylactic acid is an example of a plastic that biodegrades quickly.

Under low oxygen conditions plastics break down more slowly. The breakdown process can be accelerated in specially designed compost heap. Starch-based plastics will degrade within two to four months in a home compost bin, while polylactic acid is largely undecomposed, requiring higher temperatures. Polycaprolactone and polycaprolactone-starch composites decompose slower, but the starch content accelerates decomposition by leaving behind a porous, high surface area polycaprolactone. Nevertheless, it takes many months. In 2016, a bacterium named *Ideonella sakaiensis* was found to biodegrade PET.

Many plastic producers have gone so far even to say that their plastics are compostable, typically listing corn starch as an ingredient. However, these claims are questionable because the plastics industry operates under its own definition of compostable:

> "That which is capable of undergoing biological decomposition in a compost site such that the material is not visually distinguishable and breaks down into carbon dioxide, water, inorganic compounds and biomass at a rate consistent with known compostable materials."

The term "composting" is often used informally to describe the biodegradation of packaging materials. Legal definitions exist for compostability, the process that leads to compost. Four criteria are offered by the European Union:

- Chemical composition: Volatile matter and heavy metals as well as fluorine should be limited.
- Biodegradability: The conversion of >90% of the original material into CO_2, water and minerals by biological processes within 6 months.
- Disintegrability: At least 90% of the original mass should be decomposed into particles that are able to pass through a 2 × 2 mm sieve.
- Quality: Absence of toxic substances and other substances that impede composting.

Biodegradable Technology

Now biodegradable technology has become a highly developed market with applications in product packaging, production, and medicine. The biodegradation of biomass offers some guidances. Polyesters are known to biodegrade.

Oxo-biodegradation is defined by CEN (the European Standards Organisation) as "degradation resulting from oxidative and cell-mediated phenomena, either simultaneously or successively."

Whilst sometimes described as "oxo-fragmentable," and "oxo-degradable" these terms describe only the first or oxidative phase and should not be used for material which degrades by the process of oxo-biodegradation defined by CEN: the correct description is "oxo-biodegradable."

By combining plastic products with very large polymer molecules, which contain only carbon and hydrogen, with oxygen in the air, the product is rendered capable of decomposing in anywhere from a week to one to two years. This reaction occurs even without prodegradant additives but at a very slow rate. That is why conventional plastics, when discarded, persist for a long time in the environment. Oxo-biodegradable formulations catalyze and accelerate the biodegradation process but it takes considerable skill and experience to balance the ingredients within the formulations so as to provide the product with a useful life for a set period, followed by degradation and biodegradation.

Biodegradable technology is especially utilized by the bio-medical community. Biodegradable polymers are classified into three groups: medical, ecological, and dual application, while in terms of origin they are divided into two groups: natural and synthetic. The Clean Technology Group is exploiting the use of supercritical carbon dioxide, which under high pressure at room temperature is a solvent that can use biodegradable plastics to make polymer drug coatings. The polymer (meaning a material composed of molecules with repeating structural units that form a long chain) is used to encapsulate a drug prior to injection in the body and is based on lactic acid, a compound normally produced in the body, and is thus able to be excreted naturally. The coating is designed for controlled release over a period of time, reducing the number of injections required and maximizing the therapeutic benefit. Professor Steve Howdle states that biodegradable polymers are particularly attractive for use in drug delivery, as once introduced into the body they require no retrieval or further manipulation and are degraded into soluble, non-toxic by-products. Different polymers degrade at different rates within the body and therefore polymer selection can be tailored to achieve desired release rates.

Other biomedical applications include the use of biodegradable, elastic shape-memory polymers. Biodegradable implant materials can now be used for minimally invasive surgical procedures through degradable thermoplastic polymers. These polymers are now able to change their shape with increase of temperature, causing shape memory capabilities as well as easily degradable sutures. As a result, implants can now fit through small incisions, doctors can easily perform complex deformations, and sutures and other material aides can naturally biodegrade after a completed surgery.

Biodegradation vs. Composting

There is no universal definition for biodegradation and there are various definitions of composting, which has led to much confusion between the terms. They are often lumped together; however, they do not have the same meaning. Biodegradation is the naturally-occurring breakdown of materials by microorganisms such as bacteria and fungi or other biological activity. Composting is a human-driven process in which biodegradation occurs under a specific set of circumstances. The predominant difference between the two is that one process is naturally-occurring and one is human-driven.

Biodegradable material is capable of decomposing without an oxygen source (anaerobically) into carbon dioxide, water, and biomass, but the timeline is not very specifically defined. Similarly, compostable material breaks down into carbon dioxide, water, and biomass; however, compostable material also breaks down into inorganic compounds. The process for composting is more specifically defined, as it controlled by humans. Essentially, composting is an accelerated biodegradation process due to optimized circumstances. Additionally, the end product of composting not only returns to its previous state, but also generates and adds beneficial microorganisms to the soil called humus. This organic matter can be used in gardens and on farms to help grow healthier plants in the future. Composting more consistently occurs within a shorter time frame since it is a more defined process and is expedited by human intervention. Biodegradation can occur in different time frames under different circumstances, but is meant to occur naturally without human intervention.

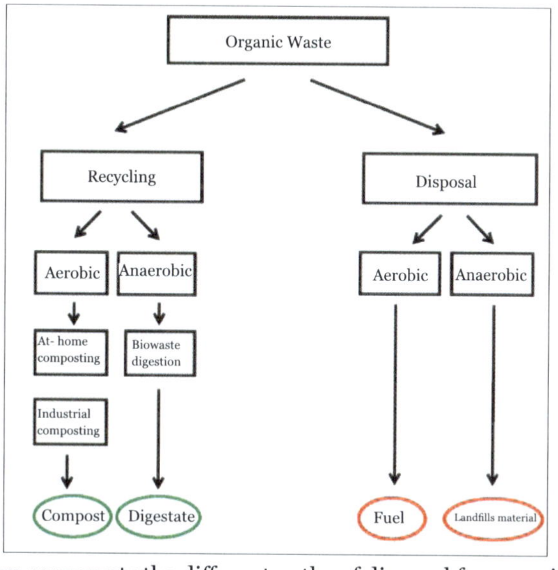

This figure represents the different paths of disposal for organic waste.

Even within composting, there are different circumstances under which this can occur. The two main types of composting are at-home versus commercial. Both produce healthy soil to be reused - the main difference lies in what materials are able to go into the process. At-home composting is mostly used for food scraps and excess garden materials, such as weeds. Commercial composting is capable of breaking down more complex plant-based products, such as corn-based plastics and larger pieces of material, like tree branches. Commercial composting begins with a manual breakdown of the materials using a grinder or other machine to initiate the process. Because at-home composting usually occurs on a smaller scale and does not involve large machinery, these materials would not fully decompose in at-home composting. Furthermore, one study has compared and contrasted home and industrial composting, concluding that there are advantages and disadvantages to both.

The following studies provide examples in which composting has been defined as a subset of biodegradation in a scientific context. The first study, "Assessment of Biodegradability of Plastics Under Simulated Composting Conditions in a Laboratory Test Setting," clearly examines composting as a set of circumstances that falls under the category of degradation. Additionally, this next study looked at the biodegradation and composting effects of chemically and physically crosslinked polylactic acid. Notably discussing composting and biodegrading as two distinct terms. The third and final study reviews European standardization of biodegradable and compostable material in the packaging industry, again using the terms separately.

The distinction between these terms is crucial because waste management confusion leads to improper disposal of materials by people on a daily basis. Biodegradation technology has led to massive improvements in how we dispose of waste; there now exist trash, recycling, and compost bins in order to optimize the disposal process. However, if these waste streams are commonly and frequently confused, then the disposal process is not at all optimized. Biodegradable and compostable materials have been developed to ensure more of human waste is able to breakdown and return to its previous state, or in the case of composting even add nutrients to the ground. When a compostable product is thrown out as opposed to composted and sent to a landfill, these inventions and efforts are wasted. Therefore, it is important for citizens to understand the difference between these terms so that materials can be disposed of properly and efficiently.

Environmental and Social Effects

Plastic pollution from illegal dumping poses health risks to wildlife. Animals often mistake plastics for food, resulting in intestinal entanglement. Slow-degrading chemicals, like polychlorinated biphenyls (PCBs), nonylphenol (NP), and pesticides also found in plastics, can release into environments and subsequently also be ingested by wildlife.

Rachel Carson, a notable environmentalist in the 1960s, provided one of the first key studies on the consequences associated with chemical ingestion in wildlife, specifically birds. In her work *Silent Spring*, she wrote on DDT, a pesticide commonly used in human agricultural activities. Birds that ate the tainted bugs, Carson found, were more likely to produce eggs with thin and weak shells.

These chemicals also play a role in human health, as consumption of tainted food (in processes called biomagnification and bioaccumulation) has been linked to issues such as cancers, neurological dysfunction, and hormonal changes. A well-known example of biomagnification impacting health in recent times is the increased exposure to dangerously high levels of mercury in fish, which can affect sex hormones in humans.

In efforts to remediate the damages done by slow-degrading plastics, detergents, metals, and other pollutants created by humans, economic costs have become a concern. Marine litter in particular is notably difficult to quanitfy and review. Researchers at the World Trade Institute estimate that cleanup initiatives' cost (specifically in ocean ecosystems) has hit close to thirteen billion dollars a year. The main concern stems from marine environments, with the biggest cleanup efforts centering around garbage patches in the ocean. In 2017, a garbage patch the size of Mexico was found in the Pacific Ocean. It is estimated to be upwards of a million square miles in size. While the patch contains more obvious examples of litter (plastic bottles, cans, and bags), tiny microplastics are nearly impossible to clean up. *National Geographic* reports that even more non-biodegradable materials are finding their way into vulnerable environments - nearly thirty-eight million pieces a year.

Materials that have not degraded can also serve as shelter for invasive species, such as tube worms and barnacles. When the ecosystem changes in response to the invasive species, resident species and the natural balance of resources, genetic diversity, and species richness is altered. These factors may support local economies in way of hunting and aquaculture, which suffer in response to the change. Similarly, coastal communities which rely heavily on ecotourism lose revenue thanks to a buildup of pollution, as their beaches or shores are no longer desirable to travelers. The World

Trade Institute also notes that the communities who often feel most of the effects of poor biodegradation are poorer countries without the means to pay for their cleanup. In a positive feedback loop effect, they in turn have trouble controlling their own pollution sources.

The first known use of *biodegradable* in a biological context was in 1959 when it was employed to describe the breakdown of material into innocuous components by microorganisms. Now *biodegradable* is commonly associated with environmentally friendly products that are part of the earth's innate cycles like the carbon cycle and capable of decomposing back into natural elements.

BIOFILTRATION

Biofiltration is a pollution control technique using a bioreactor containing living material to capture and biologically degrade pollutants. Common uses include processing waste water, capturing harmful chemicals or silt from surface runoff, and microbiotic oxidation of contaminants in air.

Examples of biofiltration include:

- Bioswales, biostrips, biobags, bioscrubbers,
- Vermifilters and trickling filters,
- Constructed wetlands and natural wetlands,
- Slow sand filters,
- Treatment ponds,
- Green belts,
- Green walls,
- Riparian zones, riparian forests, bosques,
- Bivalve bioaccumulation.

Control of Air Pollution

When applied to air filtration and purification, biofilters use microorganisms to remove air pollution. The air flows through a packed bed and the pollutant transfers into a thin biofilm on the surface of the packing material. Microorganisms, including bacteria and fungi are immobilized in the biofilm and degrade the pollutant. Trickling filters and bioscrubbers rely on a biofilm and the bacterial action in their recirculating waters.

The technology finds greatest application in treating malodorous compounds and water-soluble volatile organic compounds (VOCs). Industries employing the technology include food and animal products, off-gas from wastewater treatment facilities, pharmaceuticals, wood products manufacturing, paint and coatings application and manufacturing and resin manufacturing and application, etc. Compounds treated are typically mixed VOCs and various sulfur compounds, including hydrogen sulfide. Very large airflows may be treated and although a large area (footprint) has typically been required—a large biofilter (>200,000 acfm) may occupy as much or more land than a football field—this has been one of the principal drawbacks of the technology. Engineered biofilters, designed and built since the early 1990s, have provided significant footprint reductions over the conventional flat-bed, organic media type.

Air cycle system at biosolids composting plant.

One of the main challenges to optimum biofilter operation is maintaining proper moisture throughout the system. The air is normally humidified before it enters the bed with a watering (spray) system, humidification chamber, bioscrubber, or biotrickling filter. Properly maintained, a natural, organic packing media like peat, vegetable mulch, bark or wood chips may last for several years but engineered, combined natural organic and synthetic component packing materials will generally last much longer, up to 10 years. A number of companies offer these types or proprietary packing materials and multi-year guarantees, not usually provided with a conventional compost or wood chip bed biofilter.

Although widely employed, the scientific community is still unsure of the physical phenomena underpinning biofilter operation, and information about the microorganisms involved continues to be developed. A biofilter/bio-oxidation system is a fairly simple device to construct and operate and offers a cost-effective solution provided the pollutant is biodegradable within a moderate time frame (increasing residence time = increased size and capital costs), at reasonable concentrations (and lb/hr loading rates) and that the airstream is at an organism-viable temperature. For large volumes of air, a biofilter may be the only cost-effective solution. There is no secondary pollution (unlike the case of incineration where additional CO_2 and NO_x are produced from burning fuels) and degradation products form additional biomass, carbon dioxide and water. Media irrigation water, although many systems recycle part of it to reduce operating costs, has a moderately high biochemical oxygen demand (BOD) and may require treatment before disposal. However, this "blowdown water", necessary for proper maintenance of any bio-oxidation system, is generally accepted by municipal publicly owned treatment works without any pretreatment.

Biofilters are being utilized in Columbia Falls, Montana at Plum Creek Timber Company's fiberboard plant. The biofilters decrease the pollution emitted by the manufacturing process and the exhaust emitted is 98% clean. The newest, and largest, biofilter addition to Plum Creek cost $9.5 million, yet even though this new technology is expensive, in the long run it will cost less overtime than the alternative exhaust-cleaning incinerators fueled by natural gas (which are not as environmentally friendly).

Water Treatment

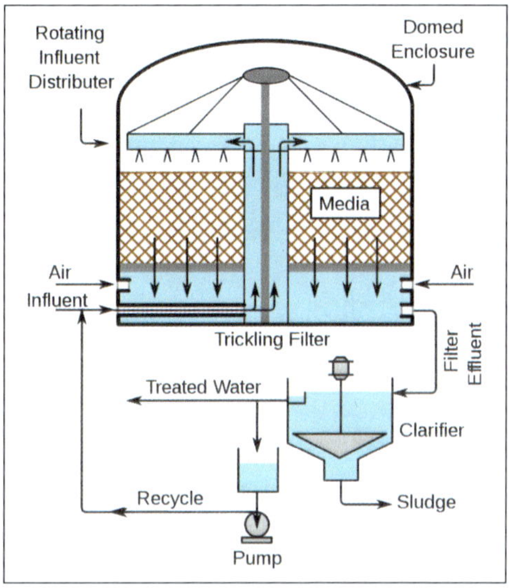

A typical complete trickling filter system for treating wastewaters.

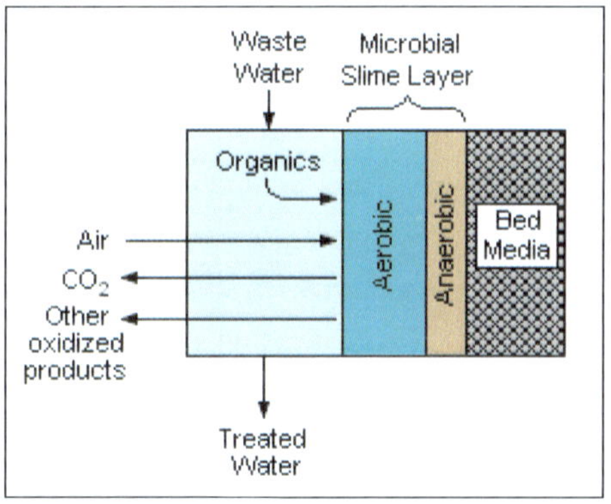

A schematic cross-section of the contact face of the bed media in a trickling filter.

Biofiltration was first introduced in England in 1893 as a trickling filter for wastewater treatment and has since been successfully used for the treatment of different types of water. Biological treatment has been used in Europe to filter surface water for drinking purposes since the early 1900s and is now receiving more interest worldwide. Biofiltration is also common in wastewater treatment, aquaculture and greywater recycling, as a way to minimize water replacement while increasing water quality.

Biofiltration Process

A biofilter is a bed of media on which microorganisms attach and grow to form a biological layer called biofilm. Biofiltration is thus usually referred to as a fixed–film process. Generally, the biofilm is formed by a community of different microorganisms (bacteria, fungi, yeast, etc.), macro-organisms (protozoa, worms, insect's larvae, etc.) and extracellular polymeric substances (EPS). The aspect of the biofilm is usually slimy and muddy.

Water to be treated can be applied intermittently or continuously over the media, via upflow or downflow. Typically, a biofilter has two or three phases, depending on the feeding strategy (percolating or submerged biofilter):

- A solid phase (media);
- A liquid phase (water);
- A gaseous phase (air).

Organic matter and other water components diffuse into the biofilm where the treatment occurs, mostly by biodegradation. Biofiltration processes are usually aerobic, which means that microorganisms require oxygen for their metabolism. Oxygen can be supplied to the biofilm, either concurrently or countercurrently with water flow. Aeration occurs passively by the natural flow of air through the process (three phase biofilter) or by forced air supplied by blowers.

Microorganisms' activity is a key-factor of the process performance. The main influencing factors are the water composition, the biofilter hydraulic loading, the type of media, the feeding strategy (percolation or submerged media), the age of the biofilm, temperature, aeration, etc.

Types of Filtering Media

Most biofilters use media such as sand, crushed rock, river gravel, or some form of plastic or ceramic material shaped as small beads and rings.

Advantages

Although biological filters have simple superficial structures, their internal hydrodynamics and the microorganisms' biology and ecology are complex and variable. These characteristics confer robustness to the process. In other words, the process has the capacity to maintain its performance or rapidly return to initial levels following a period of no flow, of intense use, toxic shocks, media backwash (high rate biofiltration processes), etc.

The structure of the biofilm protects microorganisms from difficult environmental conditions and retains the biomass inside the process, even when conditions are not optimal for its growth. Biofiltration processes offer the following advantages:

- Because microorganisms are retained within the biofilm, biofiltration allows the development of microorganisms with relatively low specific growth rates;
- Biofilters are less subject to variable or intermittent loading and to hydraulic shock;
- Operational costs are usually lower than for activated sludge;
- Final treatment result is less influenced by biomass separation since the biomass concentration at the effluent is much lower than for suspended biomass processes;
- Attached biomass becomes more specialized (higher concentration of relevant organisms) at a given point in the process train because there is no biomass return.

Drawbacks

Because filtration and growth of biomass leads to an accumulation of matter in the filtering media, this type of fixed-film process is subject to bioclogging and flow channeling. Depending on the type of application and on the media used for microbial growth, bioclogging can be controlled using physical and/or chemical methods. Whenever possible, backwash steps can be implemented using air and/or water to disrupt the biomat and recover flow. Chemicals such as oxidizing (peroxide, ozone) or biocide agents can also be used.

Drinking Water

For drinking water, biological water treatment involves the use of naturally occurring microorganisms in the surface water to improve water quality. Under optimum conditions, including relatively low turbidity and high oxygen content, the organisms break down material in the water and thus improve water quality. Slow sand filters or carbon filters are used to provide a support on which these microorganisms grow. These biological treatment systems effectively reduce water-borne diseases, dissolved organic carbon, turbidity and color in surface water, thus improving overall water quality.

Wastewater

Biofiltration is used to treat wastewater from a wide range of sources, with varying organic compositions and concentrations. Many examples of biofiltration applications are described in the literature. Bespoke biofilters have been developed and commercialized for the treatment of animal wastes, landfill leachates, dairy wastewater, domestic wastewater.

This process is versatile as it can be adapted to small flows (< 1 m3/d), such as onsite sewage as well as to flows generated by a municipality (> 240 000 m3/d). For decentralized domestic wastewater production, such as for isolated dwellings, it has been demonstrated that there are important daily, weekly and yearly fluctuations of hydraulic and organic production rates related to modern families' lifestyle. In this context, a biofilter located after a septic tank constitutes a robust process able to sustain the variability observed without compromising the treatment performance.

HYDROTHERMAL LIQUEFACTION

Hydrothermal liquefaction (HTL) is a thermal depolymerization process used to convert wet biomass into crude-like oil -sometimes referred to as bio-oil or biocrude- under moderate temperature and high pressure. The crude-like oil (or bio-oil) has high energy density with a lower heating value of 33.8-36.9 MJ/kg and 5-20 wt% oxygen and renewable chemicals.

The reaction usually involves homogeneous and/or heterogeneous catalysts to improve the quality of products and yields. Carbon and hydrogen of an organic material, such as biomass, peat or low-ranked coals (lignite) are thermo-chemically converted into hydrophobic compounds with low viscosity and high solubility. Depending on the processing conditions, the fuel can be used as produced for heavy engines, including marine and rail or upgraded to transportation fuels, such as diesel, gasoline or jet-fuels.

Chemical Reactions

In hydrothermal liquefaction processes, long carbon chain molecules in biomass are thermally cracked and oxygen is removed in the form of H_2O (dehydration) and CO_2 (decarboxylation). These reactions result in the production of high H/C ratio bio-oil. Simplified descriptions of dehydration and decarboxylation reactions can be found in the literature.

Process

Most applications of hydrothermal liquefaction operate at temperatures between 250-550 °C and high pressures of 5-25 MPa as well as catalysts for 20–60 minutes, although higher or lower temperatures can be used to optimize gas or liquid yields, respectively. At these temperatures and pressures, the water present in the biomass becomes either subcritical or supercritical, depending on the conditions, and acts as a solvent, reactant, and catalyst to facilitate the reaction of biomass to bio-oil.

The exact conversion of biomass to bio-oil is dependent on several variables:

- Feedstock composition
- Temperature and heating rate
- Pressure
- Solvent
- Residence time
- Catalysts

Feedstock

Theoretically, any biomass can be converted into bio-oil using hydrothermal liquefaction regardless of water content, and various different biomasses have been tested, from forestry and agriculture residues, sewage sludges, food process wastes, to emerging non-food biomass such as algae. The composition of cellulose, hemicellulose, protein, and lignin in the feedstock influence the yield and quality of the oil from the process.

Temperature and Heating Rate

Temperature plays a major role in the conversion of biomass to bio-oil. The temperature of the reaction determines the depolymerization of the biomass to bio-oil, as well as the repolymerization into char. While the ideal reaction temperature is dependent on the feedstock used, temperatures above ideal lead to an increase in char formation and eventually increased gas formation, while lower than ideal temperatures reduce depolymerization and overall product yields.

Similarly to temperature, the rate of heating plays a critical role in the production of the different phase streams, due to the prevalence of secondary reactions at non-optimum heating rates. Secondary reactions become dominant in heating rates that are too low, leading to the formation of char. While high heating rates are required to form liquid bio-oil, there is a threshold heating rate and temperature where liquid production is inhibited and gas production is favored in secondary reactions.

Pressure

Pressure (along with temperature) determines the super- or subcritical state of solvents as well as overall reaction kinetics and the energy inputs required to yield the desirable HTL products (oil, gas, chemicals, char etc.).

Residence Time

Hydrothermal liquefaction is a fast process, resulting in low residence times for depolymerization to occur. Typical residence times are measured in minutes (15 to 60 minutes); however, the residence time is highly dependent on the reaction conditions, including feedstock, solvent ratio and temperature. As such, optimization of the residence time is necessary to ensure a complete depolymerization without allowing further reactions to occur.

Catalysts

While water acts as a catalyst in the reaction, other catalysts can be added to the reaction vessel to optimize the conversion. Previously used catalysts include water-soluble inorganic compounds and salts, including KOH and Na_2CO_3, as well as transition metal catalysts using Ni, Pd, Pt, and Ru supported on either carbon, silica or alumina. The addition of these catalysts can lead to an oil yield increase of 20% or greater, due to the catalysts converting the protein, cellulose, and hemicellulose into oil. This ability for catalysts to convert biomaterials other than fats and oils to bio-oil allows for a wider range of feedstock to be used.

Environmental Impact

Biofuels that are produced through hydrothermal liquefaction are carbon neutral, meaning that there are no net carbon emissions produced when burning the biofuel. The plant materials used to produce bio-oils use photosynthesis to grow, and as such consume carbon dioxide from the atmosphere. The burning of the biofuels produced releases carbon dioxide into the atmosphere, but is nearly completely offset by the carbon dioxide consumed from growing the plants, resulting in a release of only 15-18 g of CO_2 per kWh of energy produced. This is substantially lower than the releases rate of fossil fuel technologies, which can range from releases of 955 g/kWh (coal), 813 g/kWh (oil), and 446 g/kWh (natural gas). Recently, Steeper Energy announced that the Carbon Intensity (CI) of its Hydrofaction™ oil is 15 CO_2eq/MJ according to GHGenius model (version 4.03a), while diesel fuel is 93.55 CO_2eq/MJ.

Hydrothermal liquefaction is a clean process that doesn't produce harmful compounds, such as ammonia, NO_x, or SO_x. Instead the heteroatoms, including nitrogen, sulfur, and chlorine, are converted into harmless byproducts such as N_2 and inorganic acids that can be neutralized with bases.

Compare with Pyrolysis and other HTL Technologies

The HTL process differs from pyrolysis as it can process wet biomass and produce a bio-oil that contains approximately twice the energy density of pyrolysis oil. Pyrolysis is a related process to HTL, but biomass must be processed and dried in order to increase the yield. The presence of water in pyrolysis drastically increases the heat of vaporization of the organic material, increasing the energy required to decompose the biomass. Typical pyrolysis processes require a water content of less than 40% to suitably convert the biomass to bio-oil. This requires considerable pretreatment of wet biomass such as tropical grasses, which contain a water content as high as 80-85%, and even further treatment for aquatic species, which can contain higher than 90% water content.

The HTL oil can contain up to 80% of the feedstock carbon content (single pass). HTL oil has good potential to yield bio-oil with "drop-in" properties that can be directly distributed in existing petroleum infrastructure.

WINDROW COMPOSTING

Windrow composting is usually considered to be the most cost-effective composting alternative. Meanwhile, it may also have the highest fugitive emissions.

Hydraulic compost turner is used in windrow composting process.

Windrow composting involves long rows of compost where the actual size and shape of the windrow are dependent on the feed stocks and type of compost turner. The piles of materials are placed in long rows, and turned by mechanical equipment to maintain optimum conditions. Mechanical turning is usually done with a full hydraulic crawler type turner machine specially designed for turning windrows. Reason of periodically turning the piles is to ensure adequate oxygen flow to the pile. Windrow composting involves a few major steps: Feedstock Preparation, Windrow Processing and, Screening.

Feedstock Preparation

- Collection: Organic waste such as green materials and food materials are recyclable through composting. Therefore, collection becomes the first step for the normal operation of industrial-scale composting facilities. Commercial and farm wastes arrive directly at the composting facility from various sources. Most of the biodegradable residuals and byproducts comes from commercial, industrial, municipal and farm/agribusiness sources. After collecting, these source-separated materials become feedstocks in a commercial composting process that turns organic waste into a valuable resource – compost.

High Moisture Fertilizer Crusher.

- Size reduction: Shredding or pulverizing reduces the size of the waste articles, resulting in a uniform mass of material. Grinding the organic material before composting greatly speeds the composting process. The smaller the size of an organic refuse particle, the more quickly the microbes can consume it. It is accomplished with semi-wet materials crusher machine. Size reduction is to provide greater surface area for microbes to attack, to improve insulation, to facilitate homogenizing of initially heterogeneous material, to reduce depth for oxygen diffusion.
- Blending: Materials can be blended and mixed to achieve a suitable C/N ratio. Microbial activity is greatest when the carbon-to-nitrogen ratio (C/N) is 30:1. This can be done easily by mixing equal parts of green and brown materials. Types and volumes of feedstocks are selected to ensure the blended admixture falls within the tolerances of each parameter, ultimately coming together to meet the desired targets for C:N ratios, moisture levels, porosity and homogeneity. In order to ensure a uniform and rapid decomposition, the compost must be thoroughly mixed.

Windrow Composting Process.

Windrow Processing

Windrow composting, employing long narrow windrows mechanically turned by a commercial compost turner on a regular basis, has become the most popular industrial composting method in the world, especially in developed countries such as United States, Australia, Italy etc.

- After crushing and blending, the material to be composted is placed into windrows, in the open air. Usually dimensions of the windrow ranging from 3-12 feed tall and 8-20 feed wide, as long as it could be. The moisture content should be 45-55 percent since poor aeration occurs at higher moisture contents and slow digestion is encountered at lower moisture contents.
- Rows are regularly turned to make sure there is enough oxygen for the microbes to work. Frequent turning is the best way to improve compost quality, to promote high concentrations of beneficial microbes, to prevent excessively high temperatures and to ensure optimum oxygen concentrations. There are many parameters affecting the frequency of windrow turning:
    - The size of the composting material and the type of composting structure also affects the frequency of turning.

Commercial compost turner in operation–windrow composting facility.

  - If C: N ratio these two conditions are at an ideal level and mixed well, expect less frequency of turning. —Turning frequency should normally be based on temperature, and should occur whenever temperatures exceed to 140 °F, or drop below 90 °F.
- It is considered finished when the organic material has fully decomposed and reached a stable state. Daily windrow composting maintains oxygenation to achieve maximum degradation, irrigation as necessary, and closely monitoring moisture content, and temperature.

Factors Affecting Composting Outdoor composting in piles that rely on mechanical aeration, typically with a compost windrow turner, to optimize the composting process. Allance compost turner machines are very efficient, which makes the whole composting process fast, high-rate, and capable of rapid degradation and stabilization of almost all organic materials.

Grinding and Screening of the Final Compost Product

The final step is high speed grinding, using chain crusher.

The screening process is to remove non-compostable material from the generated compost, and to produce a finer product that meets the requirements of consumers.

Waste is a major problem in the world organic waste amounts to almost 70% of the total waste. Another data shows that municipal solid waste currently stands at 1.3 million tons. This figure is expected to double in just over a decade. The caring for and management of global waste is a global concern for the Earth as the population of Earth grows, and places where waste is deposited becomes a problem. People are keeping looking for waste-to-energy and creative solutions for the world's waste problems, among which composting becomes an effective method to solve the the problem of organic waste — food, leaves, palm oil waste sugar press mud, biogas waste and manure etc. Commercial composting facility turns organic and biodegradable materials into compost to feed the soil at farms and residential and business landscaping projects.

Windrow Composting Advantages

- Windrow turning is important for speeding up the composting process because it breaks preferential air pathways, and brings the material from the outside to the center for pathogen kill.

- Windrow composting is a suitable option for manure management, reducing risks of odor, soil and water pollution.
- Windrow composting becomes economically and technically attractive due to its relatively simple equipment requirements and good performance results, bringing economic and environmental profits.
- Composting significantly reduces the amount of organic material ending up landfills, taking the example of Commercial composting at the Iowa City Landfill, about 7,500 tons of organics each year —yard waste, leaves and some food waste are processed into high-quality compost whose quantity arrives to 2,100 tons.
- Windrow composting does require a large amount of land and the leachate produced does have the possibility to contaminate surface and ground water, but with proper infrastructure both of those concerns are easily handled.

Application of Windrow Composting Method

- Windrow composting is suitable for large volumes of organic waste generated by entire communities and collected by local governments, and food processing businesses (e.g., restaurants, cafeterias, packing plants) and can even handle grease and animal byproducts.
- Yard waste, paper and cardboard, food waste, water treatment residuals, biodegradable plastics, pallets, sheetrock and other biodegradable materials are no problem for windrow composting process.

BIODRYING

Biodrying is the process by which biodegradable waste is rapidly heated through initial stages of composting to remove moisture from a waste stream and hence reduce its overall weight. In biodrying processes, the drying rates are augmented by biological heat in addition to forced aeration. The major portion of biological heat, naturally available through the aerobic degradation of organic matter, is utilized to evaporate surface and bound water associated with the mixed sludge. This heat generation assists in reducing the moisture content of the biomass without the need for supplementary fossil fuels, and with minimal electricity consumption. It can take as little as 8 days to dry waste in this manner. This enables reduced costs of disposal if landfill is charged on a cost per tonne basis. Biodrying may be used as part of the production process for refuse-derived fuels. Biodrying does not however greatly affect the biodegradability of the waste and hence is not stabilised. Biodried waste will still break down in a landfill to produce landfill gas and hence potentially contribute to climate change. In the UK this waste will still impact upon councils LATS allowances. Whilst biodrying is increasingly applied within commercial mechanical biological treatment (MBT) plants, it is also still subject to on-going research and development.

Process Description

In biodrying, the main drying mechanicsm is convective evaporation, using heat from the aerobic biodegradation of waste components and facilitated by the mechanically supported airflow.

The Moisture Content (MC) of the waste matrix is reduced through two main steps: (1) water molecules evaporate (i.e., change phase from liquid to gaseous) from the surface of waste fragments into the surrounding air; and (2) the evaporated water is transported through the matrix by the airflow and removed with the exhaust gasses. Limited amount of free water may seep through the waste matrix and be collected at the bottom of the biodrying reactor as leachate.

Thus in biodrying, air convection and molecular diffusion are the main transport mechanisms responsible for moisture flow through the matrix. Air convection, induced by engineered airflow through the matrix, is almost exclusively responsible for the water losses. Here, air carries the water evaporated from the surface of matrix particles (free moisture) with which is in contact.

Removal of water content from the waste matrix (desorption) by convective evaporation is governed by the thermodynamic equilibrium between the wet waste matrix (solid state) and the air flowing through the matrix (gaseous phase).

Optimal biodrying can be achieved through effective reactor design and conditioning of the input material, combined with suitable process monitoring and control. Control can be exercised by adjusting the level of operational variables (suitable to directly manipulate), informed by process state variables (suitable to monitor and evaluate). Typical design and operational choices involve:

- Matrix conditioning through mechanical pre-processing, e.g., comminution and/or mixing, affecting the physical properties of the matrix, such as the resistance to airflow;
- Type of containment of waste matrix, e.g., in enclosed boxes (or "bio-cells") or piling in tunnel windrow systems, affecting drying mechanisms including insulating effect and degree of compaction;
- Use of mixing/agitation/rotation of the waste matrix in dynamic reactors to homogenise it, i.e., achieve uniform conditions: e.g., by rotating drum reactors however, most of the existing commercial designs are static;
- Aeration system design: inverted aeration systems have been tested, intending to reduce gradients experienced in prevalent unidirectional desings;
- Management of the aeration rate of the waste matrix, by control of the inlet airflow rate (Qair), to remove water vapour and offgasses and control state process parameters, such as substrate temperature and oxygen availability;
- External systems for controlling the psychrometric properties of the inlet air (e.g., temperature, due point, relative humidity), by cooling and dehumidifying of the process air to enhance its capacity to hold water vapour, combined with partial process air recirculation;
- Residence time within the reactor, affecting the degree of completion of biochemical and physical processes. Typical residence times are in the range of 7-15 days.

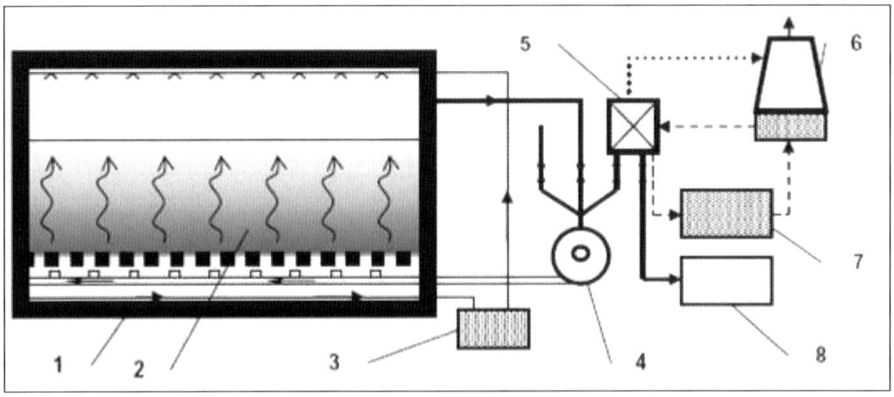

Schematic of biodrying box with process air circulation and dehumidification.

In figure, (1) enclosed box; (2) air forced through the waste matrix, heated by the exothermic aerobic biodegradation of readily decomposable waste fragments; (3) leachate collection and circulation system; (4) forced aeration system with partial air recirculation, mixing ambient air and conditioned process air; (5) heat exchanger; (6) cooling tower; (7) water (vapour condensate); (8) exhaust air treatment through biofilter or regenerative thermal oxidation (RTO). Appropriate conditions for microbial activity allow for the biodegradation of the waste placed within the bioreactor, providing the necessary heat to evaporate moisture from the waste fragments. Evaporated moisture is removed by the air convection, achieved by forced aeration. The exhaust air is going through various treatment stages that improve its drying capacity (ability to carry moisture) before it is partly re-circulated into the reactor, after being mixed with ambient air.

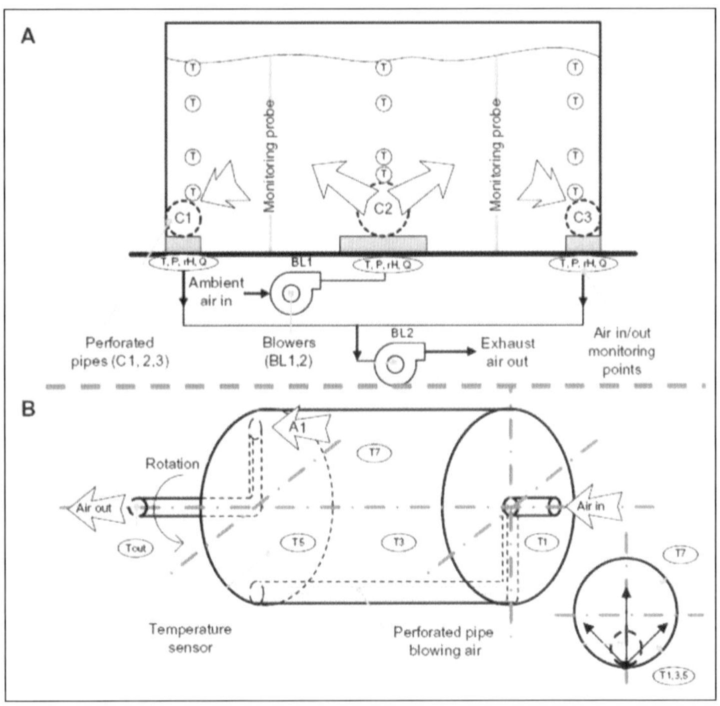

Figure: Simplified schematics of bench/pilot-scale biodrying reactor designs, among else aiming to mitigate the uneven drying of matrix. Reactor A: static enclosed hall. The central perforated pipe (C2) alternates between blowing and pulling air through the matrix, whilst the peripheral pipes (C2, C3) operate conversely. Reactor B: cylindrical rotating drum with one perforated pipe. Certain monitoring points are shown: T: temperature: 1–7 internal, out: exhaust air; P: pressure; rH: relative humidity; Q: air flowrate. BL: blower.

MC of the waste matrix is the single most important variable for evaluating the performance of biodrying processes. In waste management the MC is typically measured by gravimetric water content methods and expressed as a percentage of water for the wet weight of the material (wet basis: ar). In biodrying, the MC can be reduced from ca. 35–55% w/w to 20–10% w/w ar. During aerobic biodegradation around 0.5–0.6 g of metabolic water is produced per g of VS decomposed. However, water losses during biodrying are much greater than the gains of metabolic water, resulting in a dried matrix. Mass balance of MC should include both metabolic water gains and evaporation–convection losses. Overall weight losses of 25% w/w are considered as typical.

References

- "Composting for the Homeowner - University of Illinois Extension". Web.extension.illinois.edu. Retrieved 18 July 2013
- Bioremediation: kenyon.edu, Retrieved, 31 July, 2019
- Trautmann, Nancy. "Invertebrates of the Compost Pile". Cornell Center for the Environment. Retrieved 2012-10-03
- Biodrying, database, waste-c-control: epem.gr, Retrieved, 23 July, 2019
- Allsopp, Dennis (2004). Introduction to Biodeterioration. Cambridge: Cambridge University Press. ISBN 9780511617065
- "Pilot studies for vermifiltration of 1000m3day of sewage wastewater". Www.academia.edu. Retrieved 2016-02-21
- Navaee-Ardeh S, Bertrand F, Stuart PR (2006). "Emerging biodrying technology for the drying of pulp and paper mixed sludges". Drying Technology. 24 (7): 863–78. Doi:10.1080/07373930600734026
- Lotzof, M. "Very Large Scale Vermiculture in Sludge Stabilisation". Vermitech Pty Limited. Retrieved 2012-10-03
- Windrow-composting-technology-and-advantages, composting-technologies: compost-turner.net, Retrieved 1 August, 2019
- "Harvesting - MMSB - Multi-Materials Stewardship Board". MMSB - Multi-Materials Stewardship Board. Archived from the original on 2015-10-28. Retrieved 2016-02-21
- Pyrolysis, science: britannica.com, Retrieved 15 May, 2019

5

Waste Handling and Disposal

The process of collecting and transporting waste is termed as waste handling. There are numerous methods which are used to dispose waste such as incineration, waste compaction, biogas generation and composting. All the diverse techniques and methods of waste handling and disposal have been carefully analyzed in this chapter.

WASTE HANDLING

When we talk about solid waste management, there are many aspects of how such waste needs to be collected, treated, disposed of and other aspects. Today all modern towns and cities are facing the challenge of handling growing amounts of solid wastes and efficient management of such wastes including handling and transportation which is a subject of elaborate planning and management of the local authorities.

Waste collection varies from country to country, region to region. The collection methods depends on the expanse of areas covered, the landfill areas the garbage needs to be transported to amongst other factors. Usually the local government arranges for the waste handling and transportation in most countries though in some places it is handled by private players. Again, in many rural areas of developing countries there is no formal waste collection method.

In some advanced countries including those in Europe a vacuum based system is used to collect the refuse through underground conduits known as Envac. Here the refuse is transported to a designated area through a vacuum pump system, eliminating the need for manual handling of garbage. Metro Taifun is another vacuum based solution of single or ring line systems.

In many cities such as in Canada wastes are collected from the curbside in neighborhoods. Here the garbage bins are designated as wastes or recyclable materials and people are encouraged to segregate their waste before they drop them into the bins. These are then collected based on a daily schedule. In rural areas people need to transfer their garbage to a transfer station from where it is transported to a landfill area.

Many Asian countries face problem of huge volumes of solid waste management. In many such countries such as Taipei, the government charges taxes as per the volume of garbage produced and issue garbage bags in which the waste needs to be collected and deposited by the city people. Such initiatives have helped to reduce the waste accumulation and encouraged people to recycle and reuse.

The ArrowBio system is commonly used in many countries such as Australia, Greece, Mexico, UK or California. Here the trash is collected in trucks and by mechanical methods such as gravitational screening, settling and hydro mechanical shredding. Huge volumes of wastes are segregated, salvaged for recyclables and then turned into biogas or compost.

Thus, different countries and cities have different means of collecting wastes and transporting it but it still remains a huge responsibility on the civic authorities and they need the active help of the public in the growing waste management issues.

WASTE SORTING

Waste sorting is a simple process that requires neither a lot of time, nor any special resources. While sorting waste, the key is to separate from household waste the types of waste that can be recycled.

"Zala josta" suggests dividing the total waste stream into the following groups:

- Packaging materials (plastic, glass, paper/ cardboard);
- Metal;
- Batteries;
- Electrical equipment;
- Environmentally hazardous waste;
- Medical waste;
- Biodegradable waste;
- Reusable things;
- Bulky waste and construction waste;
- Household waste not suitable for recycling.

Plastic Packaging Waste

These are mostly empty PET bottles, containers, boxes, film and bags.

- How to collect: Prepare for recycling by washing and removing the impurities, such as food residues, flattening as much as possible.
- Suitable for recycling: Flattened plastic beverage bottles, plastic bags, film, plastic boxes.
- Not suitable for recycling: Dirty plastic products, oil bottles, yogurt, sour cream or margarine containers, toys, household goods, household items.
- Where to deliver: Any separate collection container for plastic waste (usually yellow) or to a specially arranged ECO site.

Glass Packaging Waste

They are:

- All kinds of glass jars, containers, bottles.
- How to collect: Prepare for recycling by washing and removing the impurities, such as food residues.
- Suitable for recycling: Glass jars, bottles, window glass.
- Not suitable for recycling: Mirror glass, auto glass, oil bottles, balsam bottles (of clay).
- Where to deliver: All the jars and bottles to any sortes collection container for glass waste (usually green) or a specially arranged ECO site. Window glass should be usually delivered to EKO sites.

Paper and Cardboard Wastes

They are:

- All paper packaging, cardboard, newspapers, magazines, books, note paper, egg boxes, and other paper/ cardboard products.
- How to collect: Prepare for recycling by removing the impurities and flattening as much as possible.
- Suitable for recycling: Newsprint, cardboard, writing paper, books, unsorted paper, all kinds of paper/cardboard products, also stapled.
- Not suitable for recycling: Paper and cardboard that is wet, dirty, containing food impurities are laminated.
- Where to deliver: Any separate collection container for paper waste (usually blue) or a specially arranged ECO site. Paper can also be delivered to paper-mill "Ligatne" by signing an agreement on paper waste removal.

Metal Waste

These are:

- All types of metal products, and metal packaging such as cans or tins.
- How to collect: Prepare for recycling by removing impurities such as food residues, flattening to possible extent.
- Suitable for recycling: All kinds of metal products free of contaminants.
- Not suitable for recycling: Greasy, dirty metal packages.
- Where to deliver: To any EKO site or specialized scrap purchasing points in exchange for an appropriate fee.

Waste Batteries and Accumulators

They are:

- All kinds of batteries and accumulators of various electrical devices.
- How to collect: Collected separately from other waste, keeping away from moisture.
- Suitable for recycling: Any batteries and rechargeable batteries, and batteries that have leaked.
- Where to deliver: Any specialized battery collection container, which is located in the battery POS - stores, as well as educational institutions, government bodies and specialized ECO sites.

Deteriorated Electrical Appliances

They are:

- Any (big or small) electrical and electronic devices that are operated by a power source.
- How to collect: Prepare for recycling by disconnecting from the power supply. Before scrapping it shall not be disassembled by oneself.
- Suitable for recycling: Any type of electrical equipment.
- Where to deliver: Take the opportunity to get in touch with collection companies of deteriorated electric equipment, which will remove free of charge the deteriorated appliances from your home. If you purchase new electrical equipment, according to the legislative provisions the seller shall free of charge accept from you the same kind of obsolete electrical equipment and ensure its utilization.

Environmentally Hazardous Waste

They are:

- Waste industrial oils, oil filters, used car tires.
- How to collect: Collect in specialized containers provided by the manager of environmentally hazardous products on a contract basis. Tires are collected without any special rules.
- Suitable for recycling: Environmentally hazardous waste free of contaminants.
- Where to deliver: A Specially arranged ECO site or a point of collection of environmentally hazardous products. From private customers without charge there are accepted up to 4 car tires. If you want to deliver more than 4 tires, then it is a paid service.

Medical Waste

It is any kind of medication, which has expired.

- How to collect: Collect in a separate bag. Ensure that bottles are firmly sealed.

- Suitable for disposal: Any medication, which has expired or, which is no longer used.
- Do not: Rinse medication down the sewer system, or donate unused medication to others.
- Where to deliver: Any point of collection of hazardous waste or a nearby pharmacy.

Biodegradable Waste

They are:

- Any waste of natural origin - fruit and vegetable leftovers, egg shells, biopolymers packaging, also wet paper waste.
- How to collect: Collect in a separate container preparing for composting.
- Not suitable for composting: Dairy and meat products leftovers, which during the decomposition will distribute an unpleasant odor and promote bacterial growth.
- Where to deliver: Compost at the back garden, creating a compost pile or compost box.

Reusable Things

They are:

- Clothing, shoes, household goods, textiles, furniture and other things that can be reused.
- Where to deliver: Donate to charity, the Red Cross or orphanages, homes for the elderly. Old blankets and worn clothing will serve animal shelters.

Bulky Waste and Construction Waste

They are old and broken-down shell of soft furniture, any redecoration waste.

- Where to deliver: You should contact the waste management with a request to provide you with a special container to collect the mentioned waste. This is a paid service.

Household Waste not Suitable for Recycling

Everything else that could not be classified to any of the recyclable material groups is household waste. It is brought to a regular unsorted waste container - it will be stored in a waste site and will not be recycled.

Tips on Sorting Waste

- Decline without regrets: Always refuse any material you don't need, especially packaging. Reducing the volume of trash is the best way to reduce the amount of solid waste in the landfill.

  Some stores already have places for you to leave your packaging at the time of purchase (discard your toothpaste box at the supermarket checkout, for instance).

- To wash, or not to wash, that is the question: Cleaning the leftover food from the discards will help people who work in the recycling plant. Your yogurt cup will take days or even weeks to reach the recycling site. Have you thought about the smell? Yuck! To avoid wasting water and to make the task easier, leave the recyclables in the sink so they can be washed with the water you're using for the dishes. Another approach is to take advantage of already-used napkins to clean the items.
- Compost almost everything damp or squishy: This type of waste accounts for half the volume of all solid waste, fills up landfills and produces methane gas, which is 20 times more polluting than carbon dioxide. But a lot of it can be composted. Fruit and vegetable peels, coffee grounds and even wet paper can be turned into fertilizer.

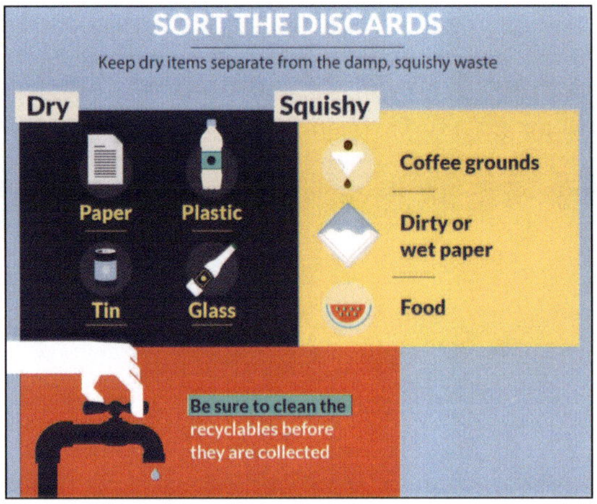

- Do you crumple up paper when throwing it in the garbage can? Stop: Avoid folding or crumpling paper before throwing it away. The reason is easy to understand: The more intact paper is, the more money it is worth, as the (cellulose) fibers are stronger, increasing the number of times it can be recycled.
- Worthless paper: On the other hand, there are many types of paper that cannot be recycled: greaseproof and plasticized paper, receipts from credit card machines or supermarkets (avoid requesting them), paper with adhesive glue (separate the glue from the paper), dirty toilet paper, carbon paper and tax coupons. (Used paper napkins and paper towels don't belong in your recycled paper bin, either, but they can be composted. So can greasy pizza boxes; that way they won't dirty the rest of the recyclable paper.)

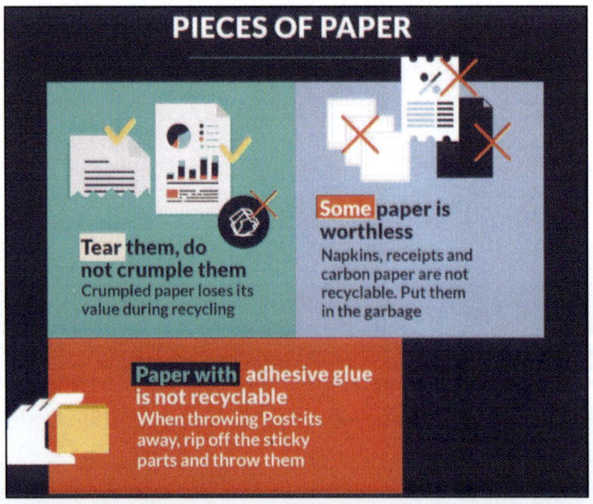

- Costly Glass: For the future, to avoid the cost of recycling glass, we must fight to increase the use of returnable containers. Until then, reuse glass to the maximum. Whenever possible, avoid long neck bottles: They are not returnable and the recycling process (which shatters them into shards) is still complex and expensive.
- Keep tabs on those tabs: Do not remove the tabs from your soda or beer cans. They can get lost amidst the refuse and end up not being recycled.

- Squeeze that toothpaste: The tube is made of low-density polyethylene, which is a type of plastic. Therefore, try to squeeze as much toothpaste from it as possible and place it in the recycling bin with the other plastic items.
- Snack packaging: You may not know this, but these products often have in their packaging a type of plastic which is covered by a thin layer of aluminum. It can be recycled. The problem is that the process is expensive, and only a few companies are able to pay for it. Find out how it works in your area. And, try to avoid buying products with this type of packaging, ok?

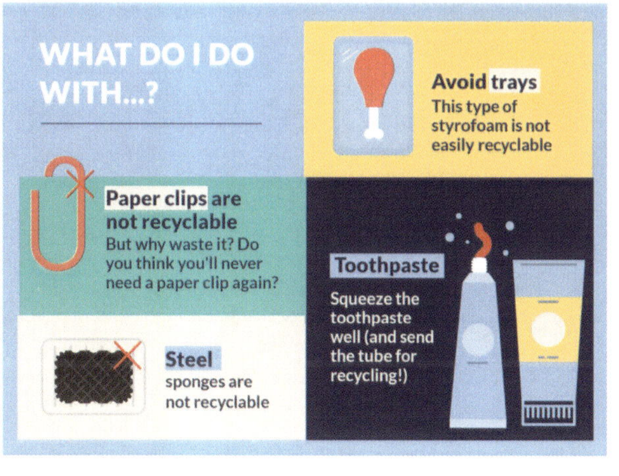

- Goodbye, polystyrene tray: When you buy electronics packed with styrofoam, recycle it. This kind of styrofoam has market value and therefore the process of recycling it is affordable. That polystyrene tray with the plastic wrap used to store fish, meat and even vegetables is another story. Because it has little market value, it is expensive to recycle. So, take containers made of glass or aluminum to your butcher or fishmonger and ask them to use these to pack your meat and fish. Fruits and vegetables do not need packaging.

- Warning: Hazardous waste cannot be thrown into the regular waste stream. Like the consumers who buy them, companies producing toxic items must take responsibility for their disposal. Thus, they must provide collection sites, where the materials can be safely picked up and recycled. Some examples are:
    - Bottles and packs of medicine: The blister packs cannot be recycled normally and should not be placed in the bin with the rest of your plastics. Like some medicine bottles, they contain remnants of chemical compounds that can contaminate the environment. They're also, in some countries, made with a highly toxic plastic that is one of the least recyclable. So, you must take them to collection sites, which can be found in many drugstores. Depending on where you live, some municipalities collect all these items at "medical collection sites," along with unused medicines.
    - Batteries, lamps and ink cartridges: There are collection sites at supermarkets and specialized stores for all of these items. In the case of the ink cartridge, it's best to let the ink dry for at least 24 hours – this makes its container easier to recycle.
    - Toxic Brew: Coffee capsules are very difficult to recycle because, like the snack packaging mentioned above, they contain a couple different materials: aluminum and plastics. Fortunately, the major brands selling these products offer collection sites at their stores. But consider doing yourself and the planet a favor and buying a coffee machine that uses grounds.

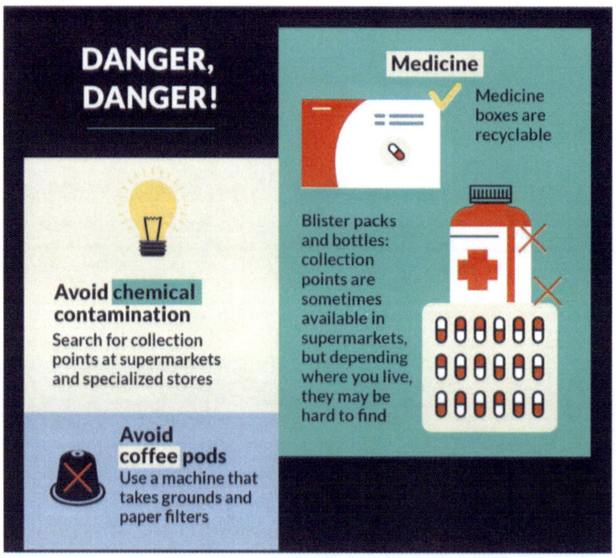

- Do you know what garbage looks like? Are you sure?: There are objects that are not recyclable, and some of them will surprise you. Besides all the non-recyclable paper mentioned above, film wrap, paper clips, steel sponges and tape are also considered garbage. They go straight to the landfill, along with the non-compostable damp garbage (dirty diapers, used kitty litter, all that lovely aromatic stuff).

WASTE CONTAINER

A waste container is a container for temporarily storing waste, and is usually made out of metal or plastic. Some common terms are dustbin, garbage can, and trash can. The words "rubbish", "basket" and "bin" are more common in British English usage; "trash" and "can" are more common in American English usage. "Garbage" may refer to food waste specifically (when distinguished from "trash") or to municipal solid waste in general. In 1875, the first household rubbish bins were introduced in Britain to create a regulated system of collection.

Public Waste Collection

In many cities and towns, there is a public waste collection service which regularly collects household waste from outside buildings etc. This will be loaded into a garbage truck and driven to a landfill, incinerator or crush facility to be disposed of. Household waste containers are usually either:

- Trash cans, receptacles made of metal or plastic.
- Wheelie bins, light, mobile plastic bins.

In some areas, each household has multiple bins for different categories of rubbish (usually represented by colours) depending on its suitability for recycling.

Art on waste containers.

Commercial roadside waste containers are often larger dumpsters or skips.

Public Litter Bins

Waste containers at the National Theater in Taipei, Taiwan

Public areas, such as parks, often have litter bins placed to improve the social environment by encouraging people not to litter. Such bins in outdoor locations or other busy public areas are usually mounted to the ground or wall to discourage theft, and reduce vandalism, and to improve their appearance are sometimes deliberately artistic or cute. In dense urban areas, trash is stored underground below the receptacle. Some trash cans have plastic bags to help contain liquids.

Metaphors

International symbol "Tidyman" used on packaging to remind people to dispose of it in a bin instead of littering.

The term "garbage can" is also used for a model of decision making, the "Garbage Can Model" of decision making. It is concerned with cases of decision making in great aggregate uncertainty which can cause decisions to arise that from a distant point of view might seem irrational. A "trash can" metaphor is often used in computer operating system desktop environments as a place files can be moved for deletion.

In a workplace setting, a bin may be euphemistically called "the circular file", "the round file" or "the janitor's file". Whereas useful documents are filed in a filing cabinet, which is rectangular, junk mail and other worthless items are "filed" in the bin, which is often round.

The term "waste basket" is occasionally used in taxonomy to refer to groupings that are based on some non-genetic criterion (e.g., the proposed order Insectivora is considered a "wastebasket taxon", as it groups small mammals whose chief link is that they eat insects), and the Nilo-Saharan language family is sometimes called "Greenberg's wastebasket", as it was a grouping made by him to fit the languages of Africa that did not fall into the other groups, Afroasiatic, Niger–Congo, and Khoisan.

LANDFILL

A landfill site (also known as a tip, dump, rubbish dump, garbage dump or dumping ground and historically as a midden) is a site for the disposal of waste materials by burial. Landfill is the oldest form of waste treatment, although the burial of the waste is modern; historically, refuse was simply left in piles or thrown into pits. Historically, landfills have been the most common method of organized waste disposal and remain so in many places around the world.

Some landfills are also used for waste management purposes, such as the temporary storage, consolidation and transfer, or processing of waste material (sorting, treatment, or recycling). Unless they are stabilized, these areas may experience severe shaking or soil liquefaction of the ground during a large earthquake.

Operations

One of several landfills used by Dryden, Ontario, Canada.

Typically, operators of well-run landfills for non-hazardous waste meet predefined specifications by applying techniques to:

- Confine waste to as small an area as possible.
- Compact waste to reduce volume.

They can also cover waste (usually daily) with layers of soil or other types of material such as woodchips and fine particles.

During landfill operations, a scale or weighbridge may weigh waste collection vehicles on arrival and personnel may inspect loads for wastes that do not accord with the landfill's waste-acceptance criteria. Afterward, the waste collection vehicles use the existing road network on their way to the tipping face or working front, where they unload their contents. After loads are deposited, compactors or bulldozers can spread and compact the waste on the working face. Before leaving the landfill boundaries, the waste collection vehicles may pass through a wheel-cleaning facility. If necessary, they return to the weighbridge for re-weighing without their load. The weighing process can assemble statistics on the daily incoming waste tonnage, which databases can retain for record keeping. In addition to trucks, some landfills may have equipment to handle railroad containers. The use of "rail-haul" permits landfills to be located at more remote sites, without the problems associated with many truck trips.

Typically, in the working face, the compacted waste is covered with soil or alternative materials daily. Alternative waste-cover materials include chipped wood or other "green waste", several sprayed-on foam products, chemically "fixed" bio-solids, and temporary blankets. Blankets can be lifted into place at night and then removed the following day prior to waste placement. The space that is occupied daily by the compacted waste and the cover material is called a daily cell. Waste compaction is critical to extending the life of the landfill. Factors such as waste compressibility, waste-layer thickness and the number of passes of the compactor over the waste affect the waste densities.

Sanitary Landfill Life Cycle

The term landfill is usually shorthand for a municipal landfill or sanitary landfill. These facilities were first introduced early in the 20th century, but gained wide use in the 1960s and '70s, in an effort to eliminate open dumps and other "unsanitary" waste disposal practices. The sanitary landfill is an engineered facility that separates and confines waste. Sanitary landfills are intended as biological reactors (bioreactors) in which microbes will break down complex organic waste into simpler, less toxic compounds over time. These reactors must be designed and operated according to regulatory standards and guidelines.

Usually, aerobic decomposition is the first stage by which wastes are broken down in a landfill. These are followed by four stages of anaerobic degradation. Usually, solid organic material in solid phase decays rapidly as larger organic molecules degrade into smaller molecules. These smaller organic molecules begin to dissolve and move to the liquid phase, followed by hydrolysis of these organic molecules, and the hydrolyzed compounds then undergo transformation and volatilization as carbon dioxide (CO_2) and methane (CH_4), with rest of the waste remaining in solid and liquid phases.

During the early phases, little material volume reaches the leachate, as the biodegradable organic matter of the waste undergoes a rapid decrease in volume. Meanwhile, the leachate's chemical oxygen demand increases with increasing concentrations of the more recalcitrant compounds compared to the more reactive compounds in the leachate. Successful conversion and stabilization of the waste depends on how well microbial populations function in syntrophy, i.e. an interaction of different populations to provide each other's nutritional needs.

The life cycle of a municipal landfill undergoes five distinct phases:

Phase I – Initial adjustment:

As the waste is placed in the landfill, the void spaces contain high volumes of molecular oxygen (O_2). With added and compacted wastes, the O_2 content of the landfill bioreactor strata gradually decreases. Microbial populations grow, density increases. Aerobic biodegradation dominates, i.e. the primary electron acceptor is O_2.

Phase II – Transition:

The O_2 is rapidly degraded by the existing microbial populations. The decreasing O_2 leads to less aerobic and more anaerobic conditions in the layers. The primary electron acceptors during transition are nitrates and sulphates, since O_2 is rapidly displaced by CO_2 in the effluent gas.

Phase III – Acid formation:

Hydrolysis of the biodegradable fraction of the solid waste begins in the acid formation phase, which leads to rapid accumulation of volatile fatty acids (VFAs) in the leachate. The increased organic acid content decreases the leachate pH from approximately 7.5 to 5.6. During this phase, the decomposition intermediate compounds like the VFAs contribute much COD. Long-chain volatile organic acids (VOAs) are converted to acetic acid ($C_2H_4O_2$), CO_2, and hydrogen gas (H_2). High concentrations of VFAs increase both the biochemical oxygen demand (BOD) and VOA concentrations, which initiates H_2 production by fermentative bacteria, which stimulates the growth of H_2-oxidizing bacteria. The H_2 generation phase is relatively short because it is complete by the end of the acid formation phase. The increase in the biomass of acidogenic bacteria increases the amount of degradation of the waste material and consuming nutrients. Metals, which are generally more water-soluble at lower pH, may become more mobile during this phase, leading to increasing metal concentrations in the leachate.

Phase IV – Methane fermentation:

The acid formation phase intermediary products (e.g. acetic, propionic, and butyric acids) are converted to CH_4 and CO_2 by methanogenic microorganisms. As VFAs are metabolized by the methanogens, the landfill water pH returns to neutrality. The leachate's organic strength, experessed as oxygen demand, decreases at a rapid rate with increases in CH_4 and CO_2 gas production. This is the longest decomposition phase.

Phase V – Final maturation and stabilization:

The rate of microbiological activity slows during the last phase of waste decomposition as the supply of nutrients limits the chemical reactions, e.g. as bioavailable phosphorus becomes increasingly scarce. CH_4 production almost completely disappears, with O_2 and oxidized species gradually reappearing in the gas wells as O_2 permeates downwardly from the troposphere. This transforms the oxidation–reduction potential (ORP) in the leachate toward oxidative processes. The residual organic materials may incrementally be converted to the gas phase, and as organic matter is composted; i.e. the organic matter is converted to humic-like compounds.

Social and Environmental Impact

Landfill operation in Hawaii. Note that the area being filled is a single, well-defined "cell" and that a protective landfill liner is in place (exposed on the left) to preventcontamination by leachates migrating downward through the underlying geological formation.

Landfills have the potential to cause a number of issues. Infrastructure disruption, such as damage to access roads by heavy vehicles, may occur. Pollution of local roads and water courses from wheels on vehicles when they leave the landfill can be significant and can be mitigated by wheel washing systems. Pollution of the local environment, such as contamination of groundwater or aquifers or soil contamination may occur, as well.

Leachate

When precipitation falls on open landfills, water percolates through the garbage and becomes contaminated with suspended and dissolved material, forming leachate. If this is not contained it can contaminate groundwater. All modern landfill sites use a combination of impermeable liners several metres thick, geologically stable sites and collection systems to contain and capture this leachate. It can then be treated and evaporated. Once a landfill site is full, it is sealed off to prevent precipitation ingress and new leachate formation. However, liners must have a lifespan, be it several hundred years or more. Eventually, any landfill liner could leak, so the ground around landfills must be tested for leachate to prevent pollutants from contaminating groundwater.

Decomposition Gases

Rotting food and other decaying organic waste creates decomposition gases, especially CO_2 and CH_4 from aerobic and anaerobic decomposition, respectively. Both processes occur simultaneously in different parts of a landfill. In addition to available O_2, the fraction of gas constituents will vary, depending on the age of landfill, type of waste, moisture content and other factors. For example, the maximum amount of landfill gas produced can be illustrated a simplified net reaction of diethyl oxalate that accounts for these simultaneous reactions:

$$C_6H_{10}O_4 + 1{:}5H_2O \rightarrow 3{:}25CH_4 + 2{:}75CO_2$$

On average, about half of the volumetric concentration of landfill gas is CH_4 and slightly less than half is CO_2. The gas also contains about 5% molecular nitrogen (N_2), less than 1% hydrogen sulfide (H_2S), and a low concentration of non-methane organic compounds (NMOC), about 2700 ppmv.

Landfill gases can seep out of the landfill and into the surrounding air and soil. Methane is a greenhouse gas, and is flammable and potentially explosive at certain concentrations, which makes it perfect for burning to generate electricity cleanly. Since decomposing plant matter and food waste only release carbon that has been captured from the atmosphere through photosynthesis, no new carbon enters the carbon cycle and the atmospheric concentration of CO_2 is not affected. Carbon dioxide traps heat in the atmosphere, contributing to climate change. In properly managed landfills, gas is collected and flared or recovered for landfill gas utilization.

Vectors

Poorly run landfills may become nuisances because of vectors such as rats and flies which can spread infectious diseases. The occurrence of such vectors can be mitigated through the use of daily cover.

Other Nuisances

Other potential issues include wildlife disruption, dust, odor, noise pollution, and reduced local property values.

Landfill Gas

Gases are produced in landfills due to the anaerobic digestion by microbes. In a properly managed landfill this gas is collected and used. Its uses range from simple flaring to the landfill gas utilization and generation of electricity. Landfill gas monitoring alerts workers to the presence of a build-up of gases to a harmful level. In some countries, landfill gas recovery is extensive; in the United States, for example, more than 850 landfills have active landfill gas recovery systems.

Regional Practice

A landfill in Perth, Western Australia.

South East New Territories Landfill, Hong Kong.

Canada

Landfills in Canada are regulated by provincial environmental agencies and environmental protection legislation. Older facilities tend to fall under current standards and are monitored for leaching. Some former locations have been converted to parkland.

European Union

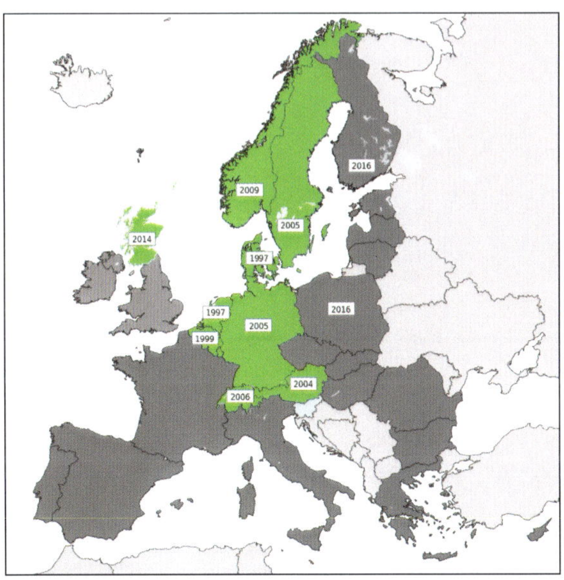

Landfill bans in Europe.

In the European Union, individual states are obliged to enact legislation to comply with the requirements and obligations of the European Landfill Directive. In the UK this is the Waste Implementation Programme. Finland, Sweden, Denmark, Poland, Germany, the Netherlands, Belgium, Austria and Slovenia have laws banning or severely restricting the disposal of household trash via landfills.

United Kingdom

Landfilling practices in the UK have had to change in recent years to meet the challenges of the European Landfill Directive. The UK now imposes landfill tax upon biodegradable waste which is put into landfills. In addition to this the Landfill Allowance Trading Scheme has been established for local authorities to trade landfill quotas in England. A different system operates in Wales where authorities are not able to 'trade' between themselves, but have allowances known as the Landfill Allowance Scheme.

United States

U.S. landfills are regulated by each state's environmental agency, which establishes minimum guidelines; however, none of these standards may fall below those set by the United States Environmental Protection Agency (EPA).

Permitting a landfill generally takes between five and seven years, costs millions of dollars and requires rigorous siting, engineering and environmental studies and demonstrations to ensure local environmental and safety concerns are satisfied.

Types

- Municipal solid waste: takes in household waste and nonhazardous material. Included in this type of landfill is a Bioreactor Landfill that specifically degrades organic material.
- Industrial waste: for commercial and industrial waste. Other related landfills include Construction and Demolition Debris Landfills and Coal Combustion Residual Landfills.
- Hazardous waste or PCB waste: Polychlorinated Biphenyl (PCB) landfills that are monitored in the United States by the Toxic Substances Control Act of 1976 (TSCA).

Microbial Topics

The status of a landfill's microbial community may determine its digestive efficiency.

Bacteria that digest plastic have been found in landfills.

Reclaiming Materials

Landfills can be regarded as a viable and abundant source of materials and energy. In the developing world, waste pickers often scavenge for still-usable materials. In a commercial context, landfill sites have also been discovered by companies, and many have begun harvesting materials and energy. Other commercial facilities include waste incinerators which have built-in material recovery. This material recovery is possible through the use of filters (electro filter, active carbon and potassium filter, quench, HCl-washer, SO_2-washer, bottom ash-grating, etc.).

Alternatives

In addition to waste reduction and recycling strategies, there are various alternatives to landfills, including waste-to-energy incineration, anaerobic digestion, composting, mechanical biological treatment, pyrolysis and plasma arc gasification. Depending on local economics and incentives, these can be made more financially attractive than landfills.

Restrictions

European countries that have banned the disposal of untreated waste in landfills.

Countries including Germany, Austria, Sweden, Denmark, Belgium, the Netherlands, and Switzerland, have banned the disposal of untreated waste in landfills. In these countries, only certain hazardous wastes, fly ashes from incineration or the stabilized output of mechanical biological treatment plants may still be deposited.

GARDEN WASTE DUMPING

Garden waste, or green waste dumping is the act of discarding or depositing garden waste somewhere it does not belong.

Garden waste is the accumulated plant matter from gardening activities which involve cutting or removing vegetation, i.e. cutting the lawn, weed removal, hedge trimming or pruning consisting of lawn clippings. leaf matter, wood and soil.

The composition and volume of garden waste can vary from season to season and location to location. A study in Aarhus, Denmark, found that on average, garden waste generation per person ranged between 122 kg to 155 kg per year.

Garden waste may be used to create compost or mulch, which can be used as a soil conditioner, adding valuable nutrients and building humus. The creation of compost requires a balance between, nitrogen, carbon, moisture and oxygen. Without the ideal balance, plant matter may take a long time to break down, drawing nitrogen from other sources, reducing nitrogen availability to existing vegetation which requires it for growth.

The risk of dumping garden waste is that it may contain seeds and plant parts that may grow (propagules), as well as increase fire fuel loads, disrupt visual amenity, accrue economic costs associated with the removal of waste as well as costs associated with the mitigation of associated impacts such as weed control, forest fire.

Cause

There are strong links between weed invasion of natural areas and the proximity and density of housing. The size and duration of the community have a direct relation to the density of weed infestation. Of the various means in which migration of exotic species from gardens take place, such as vegetative dispersal of runners, wind born and fallen seed, garden waste dumping can play a significant role. The results of one North German study found that of the problematic population of Fallopia, app. 29% originated from garden waste. Of a population of Heracleum mantegazzianum, 18% was found by Schepker to be generated by garden waste.

An Australia government publication suggest that some of the main reasons for the dumping of garden waste can be attributed to lack of care for the environment, convenience, or a reluctance to pay for the correct collection or disposal of the waste. People dump garden waste to avoid disposal fees at landfill sites or because they do not want to spend the time or effort disposing of or recycling their waste properly. This activity is carried out by people in all parts of the community, from householders to businesses, such as professional landscapers and gardeners. The spread of exotic vegetation can out-compete locally endemic vegetation, altering the composition and structure of an ecosystem.

Dumping of garden waste in particular facilitates the spread of exotic vegetation into forest remnants via the introduction of seeds and propagules contained within the garden waste. Common selection criteria for home gardeners when choosing plants are often based on ease of propagation, suitability to local environmental conditions and novelty. These specific chosen characteristics increase the chance of plant parts and seeds that are introduced into forested areas becoming a problem.

The three major causes of animal habitat degradation are; the disruption or loss of ecosystem functions, the loss of food resources and loss of resident species. Non-native invaders can cause extinctions of vulnerable native species through competition, pest and disease transportation and habitat and ecosystem alteration.

The dumping of garden waste in nature reserves surrounding and near urban areas increases the risk of fires. The dumped garden waste will eventually dry out creating fuel adding to already fallen debris fuel load on which a fire can thrive and spread on. Garden waste can spread weeds and these weeds build fuel for fires. Dumped garden waste can facilitate higher rates of erosion by smothering natural vegetation cover. With no root systems for stabilisation the top soil is vulnerable to erosion. This can add higher levels of sediments, contributing to the siltation of creeks and waterways.

If plant matter gets into waterways it can create reduced oxygen levels through the process of decomposition of green waste such as lawn clippings. This directly upsets the quality of water, affecting fish and aquatic wildlife. This dumping of green waste can also lead to the blocking of drainage systems; directly through the build-up of plant debris, and indirectly through the spread of invasive plant species that colonise wet areas, reducing and or changing the flow of waterways. This change in flow, including path and velocity, can alter hydrological cycles, affecting frequency and intensity of floods.

Impact

Green and garden waste has a direct impact to the visual aesthetics of land and can often attract further illegal dumping.

Increased Fire Risk

Dumping garden waste in nature reserves and parks surrounding and near urban areas can effect directly and indirectly the existing flora and fauna, as well as human life through the increased risk of fires. The dumped garden waste will eventually dry, creating additional fuel, adding to already fallen debris on which a fire can thrive and spread. Garden waste can spread weeds and these weeds also build fuel for fires. Fires may also spread to the suburban areas where humans can also be impacted by losing their homes from fire, incur injury or death from smoke or burns, and suffer economic loses such as income loss and clean-up costs. Fires can lead to an overall loss of habitat and biodiversity.

Threat to Biodiversity

The invasion of exotic plant species into remnant native forests is a threat to biodiversity. Some impacts of habitat degradation include; when native animals, insects and birds become vulnerable and put at risk; loss of food source for native wildlife; disruption of native plant-animal relationships ie pollination and seed dispersal and disconnection of plant-host relationships. Highly adaptive plants chosen for their ease of cultivation out compete more specialised species. Weed invasion of a forest system can change the processes of plant succession (the system of one species replacing another due to disturbance factors), the composition of the plant community and the composition and availability of nutrients. The change in forest composition can lead to loss of unique plant species. When a habitat is destroyed, the plants, animals, and other organisms that occupied the habitat have a reduced carrying capacity so that populations decline and extinction becomes a threat. Many endemic organisms have very specific requirements for their survival that can only be found within a certain ecosystem. The term 'hotspot' is used to describe areas featuring exceptional concentrations of endemic species and facing high potential of habitat degradation. The 25 most significant hotspots contain the habitats of 133,149 plant species and 9,645 vertebrate species. These endemics are confined to an expanse of 2.1 million square kilometers (1.4% of land surface). Having lost 88% of their primary vegetation, they formerly occupied 17.4 million square kilometers or 11.8% of land surface. The recruitment of alien invasive species may lead to a homogenisation of landscapes. Although increased bio diversity in subregions created by newly introduced species may occur, the displacement of the existing plant species may lead to reduced biodiversity on a global scale.

When population-level properties that indicate superior competitive ability of the invading species are examined, (42–77%) of the species are included, with the majority of species showing traits capable of modifying natural systems at both ecosystem and community/population scales.

Waterways Quality

The dumping of green waste such as lawn clippings can block drainage systems which pollutes water therefore effecting water quality and the health of aquatic plants and animals. Dumped garden waste can add high levels of sediments, reducing the light available for photosynthesis. Dumping also block waterways and roads, cause flooding and facilitate higher rates of erosion by smothering natural vegetation cover.

Causes/Stakeholders

Illegal dumping is carried out by all types of people in all parts of the community, from householders to businesses and other organizations. Addressing these motivations will enable strategies to be developed that deal with the root causes, rather than the results, of illegal dumping.

Some of the main reasons for this careless disregard for waste can be put down to sheer convenience, lack of care for the environment and also a reluctance to pay for the correct collection or disposal of the waste. The monitoring of illegally dumped garden waste by the community and industries will drive effectual tactics to battle illegal depositing. People dump waste illegally to avoid disposal fees at landfill sites or because they do not want to spend the time or effort disposing of or recycling their waste properly. Alligator weed (*Alternanthera philoxeroides* (Mart.) Griseb.) is an introduced weed originating from Sri Lanka and is creating major issues throughout the Australia since its introduction into the country. Alligator weed has the potential to affect aquatic and terrestrial biodiversity severely and to cause considerable social and economic costs, particularly in aquatic situations.

Mitigation

Education on the value of biodiversity and the negative effects of weed invasion have been identified as key items in changing current trends. Specific education campaigns on the risks of dumping garden waste could be targeted at high-risk societal groups such as residents of housing in close proximity to reserves as well as members of gardening communities and plant sellers. Restricting the selection of garden species in new housing developments adjacent to reserves may reduce the effects of illegal dumping, thereby reducing requirement and associated cost of weed management. Creating habitat for wildlife by planting native plants, making a water source available, provide shelter and places to raise young. Healthy ecosystems are necessary for the survival and health of all organisms, and there are a number of ways to reduce negative impact on the environment. Cultivation of native plant species may benefit not only native plant populations but also native animal populations. For example, Sears & Anderson suggest that native bird species diversity in Australia and North America tend to match the volume and diversity of native vegetation. Crisp also explains the percentage of native insect species in a fauna has been found to be consistent with the percentage of native plant species.

Composting is a great way to recycle nutrients back into soils. Mulching the garden with leaves and clippings. Fostering an appreciation of local natural environmental features and plant species may also help mitigate the issue. as well as the restriction of highly invasive plant species through international policy.

Utilization of green waste bins that are provided by some councils or shires that are emptied via curbside collection. The addition of facilities for waste disposal could also improve the issue. Mitigation may involve governments holding campaigns that show people disposing legally and reporting the consequences for disposing illegally. A way Australian governments are addressing the problem is through the increase of fines in conjunction with better law enforcement. In Australia, fines can be up to $1,000,000 and can also incur imprisonment. The Protection of the Environment Operations Act imposes penalties for offences including polluting waters with waste, polluting land, illegally dumping waste or using land as an illegal waste facility.

GARBAGE TRUCK

Garbage truck or dustcart refers to a truck specially designed to collect municipal solid waste and haul the collected waste to a solid waste treatment facility such as a landfill. Other common names for this type of truck include trash truck in the United States, and rubbish truck, junk truck, dumpster, bin wagon, dustbin lorry, bin lorry or bin van elsewhere. Technical names include waste collection vehicle and refuse collection vehicle. These trucks are a common sight in most urban areas.

Types

Front Loaders

A standard Waste Management Inc. front-loading garbage truck in San Jose.

Front loaders generally service commercial and industrial businesses using large waste containers with lids known as Dumpsters in the US. The truck is equipped with powered forks on the front which the driver carefully aligns with sleeves on the waste container using a joystick or a set of levers. The waste container is then lifted over the truck. Once it gets to the top the container is then flipped upside down and the waste or recyclable material is emptied into the vehicle's hopper. Once the waste is dumped, it is compacted by a hydraulically powered moving wall that oscillates backwards and forwards to push the waste to the rear of the vehicle.

Most of the newer packing trucks have "pack-on-the-go hydraulics" which lets the driver pack loads while driving, allowing faster route times. When the body is full, the compaction wall moves all the way to the rear of the body, ejecting it via an open tailgate. There is also a system called the

Curotto Can which is an attachment for a front loader that has an automated arm that functions as an automated side loader that allows the driver to dump carts.

14.5 m^3 rear load container serviced in Copenhagen.

Garbagemen loading garbage by hand in Japan.

Rear Loaders

Rear loaders have an opening into a trough or hopper at the rear that a waste collector can throw waste bags or empty the contents of bins into. Often in many areas they have a lifting mechanism to automatically empty large carts without the operator having to lift the waste by hand.

Another popular system for the rear loader is a rear load container specially built to fit a groove in the truck. The truck will have a chain or cable system for upending the container. The waste will then slide into the hopper of the truck.

The modern rear loader usually compacts the waste using a hydraulically powered mechanism that employs a moving plate or shovel to scoop the waste out from the loading hopper and compress it against a moving wall. In most compactor designs, the plate has a pointed edge (hence giving it the industry standard name *packer blade*) which is designed to apply point pressure to the waste to break down bulky items in the hopper before being drawn into the main body of the truck.

Compactor designs, however, have been many and varied, however the two most popular in use today are the "sweep and slide" system (first pioneered on the Leach 2R Packmaster), where the packer blade pivots on a moving carriage which slides back and forth, and the "swing link" system (such as the Dempster Routechief) where the blade literally swings on a "pendulum"-style mechanism. The Heil Colectomatic used a combination of a lifting loading hopper and a pivoting sweeper blade to clear and compact waste in anticipation of the next load.

So-called "continuous" compactors were popular in the 1960s and 1970s. The German Shark design (later Rotopress) used a huge rotating drum, analogous to a cement mixer, in conjunction with a serrated auger to grind down and compact the garbage. SEMAT-Rey of France pioneered the rotating rake system (also used in the British Shelvoke and Drewry Revopak) to both mutilate waste and break down large items. Other systems used a large Archimedes' screw to draw in waste and mutilate it inside the body. A mixture of safety concerns, and higher fuel consumption has seen a decline in the popularity of continuously compacting garbage trucks. The Rotopress design remains popular due to its niche in being able to effectively deal with green waste for composting.

The wall will move towards the front of the vehicle as the pressure forces the hydraulic valves to open, or as the operator moves it with a manual control.

A unique rear-loading system involves a rear loader and a front-loading tractor (usually a Caterpillar front loader with a Tink Claw) for yard waste collection (and in some cities, garbage and recycling). The front loader picks up yard waste set in the street, and then loaded into the back of a rear loader. This system is used in several cities, including San Jose.

Side Loaders

Side loaders are loaded from the side, either manually, or with the assistance of a joystick-controlled robotic arm with a claw, used to automatically lift and tip wheeled bins into the truck's hopper.

Manual Side Loaders

A Lōdal Evo T-28 manual side loader

Manual side loaders (MSLs) feature a hopper in front of the body, similar to front loaders. Unlike front loaders, the actual hopper is very short, and sometimes is lower than the body, in order for the operator to dump the waste into the hopper. This also results in a longer times for packing loads compared to rear loaders, although this can be offset by the time bringing waste to the truck. Even though this is not only always the case, the hopper can sometimes be accessed from both sides, allowing two persons to collect waste from both waste, cutting down on the time spent doing so.

Automated Side Loaders

Lift-equipped trucks are referred to as automated side loaders (ASLs). Similar to a front-end loader, the waste is compacted by an oscillating packer plate at the front of the loading hopper which forces the waste through an aperture into the main body and is therefore compacted towards the rear of the truck.

A WhiteGMC WXLL/Heil Python automated side loader.

An automated side loader only needs one operator, whereas a traditional rear load garbage truck may require two or three people, and has the additional advantage of reducing on the job injuries due to repetitive heavy lifting. Due to these advantages, ASLs have become more popular than traditional manual collection. Typically an automated side loader uses standardized wheeled carts compatible with the truck's automated lift.

As with front loaders, the compaction mechanism comprises a metal pusher plate in the collection hopper which oscillates backwards and forwards under hydraulic pressure, pushing the refuse through an aperture, thus compacting it against the material already loaded. On some ASLs there is also a "folding" crusher plate positioned above the opening in the hopper, that folds down to crush bulky items within reach of the metal pusher plate. Another compactor design is the "paddle packer" which uses a paddle that rotates from side to side, forcing refuse into the body of the truck.

Manual/Automated Side Loaders

Manual/Automated side loaders (M/ASLs), are traditional MSLs equipped with a arm for automated collection, as well as continuously running packers. This allows for functionally identical to that of an ASL, while allowing for manual dumping of waste into hopper in instances where automated collection is not feasible, such as the collection of oversized items. In addition, M/ASLs provide a cheaper upgrade path for those who wish to keep preexisting MSLs for automated collection without paying for newer and more expensive ASLs.

Semi-automated Side Loaders

An automated side loader garbage truck in Canberra, Australia.

Semi-automated side loaders are MSLs that are equipped with an automated mechanism to lift and dump manually aligned waste containers into the hopper. The primary difference between semi-automated side loaders and ASLs is that while they still only need one person to operate, he or she must exit the cab to manually bring and align containers to the loading hopper on the side of the truck and dump them.

Automated garbage collection in Aardenburg, Netherlands.

Automated side loader in operation on an Autocar truck chassis.

Garbage collection by an automatic side loader during autumn in Kelowna, British Columbia, Canada.

Pneumatic Collection

Volvo pneumatic collector used for "waste suction".

Pneumatic collection trucks have a crane with a tube and a mouthpiece that fits in a hole, usually hidden under a plate under the street. From here it will suck up waste from an underground installation. The system usually allows the driver to "pick up" the waste, even if the access is blocked by cars, snow or other barriers.

Grapple truck.

Grapple Trucks

Grapple trucks enable the collection of bulk waste. A large percentage of items in the solid waste stream are too large or too heavy to be safely lifted by hand into traditional garbage trucks. These items (furniture, large appliances, branches, logs) are called bulky waste or "oversized." The preferred method for collecting these items is with a grapple truck. Grapple trucks have hydraulic knucklebooms, tipped with a clamshell bucket, and usually include a dump body or trailer.

Roll-offs

Roll-offs are characterized by a rectangular footprint, utilizing wheels to facilitate rolling the dumpster in place. The container is designed to be transported by special roll-off trucks. They are relatively efficient for bulk loads of waste or extremely heavy loads of construction or demolition debris.

Bin Tipper

A bin tipper is a machine which mechanically lifts and inverts bins for the purpose of emptying them. They are often components of larger machines such as garbage trucks, or can be 'standalone' or mobile units. Bin tippers typically have a steel frame, guarding and cradle, with a motor or crank-handle driving a lifting mechanism, which may be hydraulic or chain operated. Bins are placed into the machine, then lifted and inverted over the destination receptacle, allowing the contents to be emptied by gravity.

A side-load bin tipper was fitted to a garbage truck as early as 1929, by the Heil company in America. In the 1950s the Dempster Dumpmaster popularized the front-end loader variant, with bins being tipped over the cab of the truck. Both types of integrated bin tipper are now common on municipal refuse collection trucks. Standalone bin tippers developed later, with the release of a machine called the Simpro Ezi-Dump in 1990.

The use of bin tippers and other lifting aids has been stimulated in recent years by research linking heavy manual lifting with musculoskeletal disorders; some government organisations, schools and companies now prohibit emptying bins by hand. Health and safety concerns have also driven the adoption of bin tippers in the manufacturing, food-processing and construction industries.

DISPOSAL METHODS

We observe heaps of garbage lying along the roads while passing through a highway. Open dumping is the most common method of waste disposal in India. The trash heaps are usually left open to the environment and the elements. These seldom have a sparse covering which can often attract pests or vermin. Sometimes, these dumps are subjected to open burning, which can release toxic fumes and smokes. There have also been instances where enough heat have been generated to trigger a spontaneous combustion. Sometimes, wastes are illegally dumped into rivers and canals or used to fill land depressions without proper consultations. These practices cause a lot of problems in the long run. These can range from the degradation of the soil quality to leaching toxic chemicals into underground water sources. Therefore, to prevent such scenarios, proper waste disposal methods should be adopted.

Garbage accumulation has never been much of a concern in the past, but due to globalization and industrialization, there is a need for a more efficient waste disposal method. Following are some of the methods that are used today.

Landfill

In this process, the waste that cannot be reused or recycled are separated out and spread as a thin layer in low-lying areas across a city. A layer of soil is added after each layer of garbage. However, once this process is complete, the area is declared unfit for construction of buildings for the next 20 years. Instead, it can only be used as a playground or a park.

WASTE RECYCLING

The definition of recycling is to pass a substance through a system that enables that substance to be reused. Waste recycling involves the collection of waste materials and the separation and clean-up of those materials. Recycling waste means that fewer new products and consumables need to be produced, saving raw materials and reducing energy consumption.

In the UK, the household and commercial sectors have relatively low recycling rates. This is in comparison to some other wastes, such as construction and demolition waste and sewage sludge. The Government is hoping to increase the amount of household waste that we recycle to 33% by 2015. Some of the materials that we can recycle include paper, plastics, metals (such as aluminium cans) and tyres.

The paper industry generates vast quantities of waste in the form of paper off-cuttings and damaged paper rolls. This paper can be put back into the pulping process and recycled. Paper recycling in the UK became popular during the 1990s. Nearly a million tonnes of paper from household waste is now recycled each year. Although paper makes up over one third of all household waste recycled, this is still no more than about 10% of the total paper consumed. In contrast, over 50% of paper waste paper produced by the newspaper industry is currently being recycled. To encourage the public to recycle waste paper, many council have arranged house to house collection schemes.

Separate bins and containers are provided specifically for paper. They are collected at regular intervals and taken to be recycled. Other recycling depots for paper can be found at municipal centres and supermarkets.

Approximately 6 to 8% of UK household waste comprises of glass jars and bottles. However, the largest producers of waste glass bottles are hotels and pubs, as the vast majority of drinks are bottled. A large proportion of glass is collected in bottle banks and taken to be recycled. There are over 20,000 bottle banks in the UK, and they are mainly found in car parks and at supermarkets. There are usually three bottle banks, one for each colour of glass: clear, green and brown. The UK currently recycles about one third of its glass. This is far behind glass recycling rates in other European countries. Switzerland and the Netherlands for example have recycling rates as high as 80%.

Plastics make up a large amount of waste, since they are available in numerous forms. There are two main types of plastic: thermoplastics, which are the most common; and thermosetts. Thermoplastics melt when heated and can therefore be remoulded. This enables thermoplastics to be recycled relatively easily. In Western Europe the largest amounts of plastic occur in the form of packaging. Plastic waste tends to be sorted by hand, either at a materials recycling facility or the householder can separate it. This may then be taken to a plastic recycling point or collected by the council. The UK produces approximately about 4.5 million tonnes of plastic waste each year. Most of this waste arises from packaging. The UK has a plastics recycling rate of only 3%. In Germany the recycling rate for plastic is 70%.

The UK has a recycling rate of approximately 60% for iron and steel. Most of this waste comes from scrap vehicles, cooker, fridges and other kitchen appliances. It is estimated that the metal content of household waste is between 5 and 10%. It is mainly made up of aluminium drinks cans and tin-plated steel food cans. Aluminium recycling is widely established in the UK. It is an expensive metal and can therefore produce high incomes for recycling schemes. Copper, zinc and lead are also recycled in the UK. At present, over a third of aluminium drinks cans are recycled. Some other countries have very high recycling figures for aluminium drinks cans. The USA and Australia for example, recycle nearly two thirds.

Every year in the UK between 25 and 30 million scrap tyres are generated. Approximately 21% of these tyres are retreaded and reused. The old tread is ground off the tyre and replaced with a new tread. However, about half of all used tyres are dumped in landfill sites throughout the country. Other tyres may be incinerated.

References

- Waste-handling-and-transport: wastemanagement.in, Retrieved 31 July, 2019
- Shendruk, amanada (5 august 2018). "could nyc solve its trash problem with underground trash cans?". Quartz (publication). Retrieved 5 september 2018
- How-sort-waste-properly: zalajosta.lv, Retrieved, 28 April, 2019
- Geroux, zachary; voytko, eric. "the ever expanding history of the front load refuse truck". Retrieved 10 september 2014
- Waste-recycling, sustainability: enviropedia.org.uk, Retrieved 18 May, 2019
- Waste-disposal, biology: byjus.com, Retrieved, 15 January, 2019

Permissions

All chapters in this book are published with permission under the Creative Commons Attribution Share Alike License or equivalent. Every chapter published in this book has been scrutinized by our experts. Their significance has been extensively debated. The topics covered herein carry significant information for a comprehensive understanding. They may even be implemented as practical applications or may be referred to as a beginning point for further studies.

We would like to thank the editorial team for lending their expertise to make the book truly unique. They have played a crucial role in the development of this book. Without their invaluable contributions this book wouldn't have been possible. They have made vital efforts to compile up to date information on the varied aspects of this subject to make this book a valuable addition to the collection of many professionals and students.

This book was conceptualized with the vision of imparting up-to-date and integrated information in this field. To ensure the same, a matchless editorial board was set up. Every individual on the board went through rigorous rounds of assessment to prove their worth. After which they invested a large part of their time researching and compiling the most relevant data for our readers.

The editorial board has been involved in producing this book since its inception. They have spent rigorous hours researching and exploring the diverse topics which have resulted in the successful publishing of this book. They have passed on their knowledge of decades through this book. To expedite this challenging task, the publisher supported the team at every step. A small team of assistant editors was also appointed to further simplify the editing procedure and attain best results for the readers.

Apart from the editorial board, the designing team has also invested a significant amount of their time in understanding the subject and creating the most relevant covers. They scrutinized every image to scout for the most suitable representation of the subject and create an appropriate cover for the book.

The publishing team has been an ardent support to the editorial, designing and production team. Their endless efforts to recruit the best for this project, has resulted in the accomplishment of this book. They are a veteran in the field of academics and their pool of knowledge is as vast as their experience in printing. Their expertise and guidance has proved useful at every step. Their uncompromising quality standards have made this book an exceptional effort. Their encouragement from time to time has been an inspiration for everyone.

The publisher and the editorial board hope that this book will prove to be a valuable piece of knowledge for students, practitioners and scholars across the globe.

Index